SpringerBriefs in Mathematics

T0281376

SpringerBriefs in Mathematics showcases expositions in all areas of mathematics and applied mathematics. Manuscripts presenting new results or a single new result in a classical field, new field, or an emerging topic, applications, or bridges between new results and already published works, are encouraged. The series is intended for mathematicians and applied mathematicians.

For further volumes:
http://www.springer.com/series/10030

Ignacio M. Pelayo

Geodesic Convexity in Graphs

 Springer

Ignacio M. Pelayo
School of Agricultural Engineering
   of Barcelona
Barcelona, Spain

ISSN 2191-8198   ISSN 2191-8201 (electronic)
ISBN 978-1-4614-8698-5   ISBN 978-1-4614-8699-2 (eBook)
DOI 10.1007/978-1-4614-8699-2
Springer New York Heidelberg Dordrecht London

Library of Congress Control Number: 2013946414

Mathematics Subject Classification (2010): 05C12, 05C35

Printed on acid-free paper

Springer is part of Springer Science+Business Media (www.springer.com)

Preface

This monograph is intended to present the current state of the art of the so-called theory of *geodesic convexity* in finite, simple, connected graphs. It has been designed with the objective of being useful and stimulating for research workers in the field and for developing courses on convexity in graphs for graduate students as well as advanced undergraduates.

Convexity, in both continuous and discrete structures, has been studied in many contexts. These contexts have been generalized to the concept of a *convexity space*, which is a pair (V,C), where V is a set and C is a collection of subsets of V, called *convex sets*, such that $\emptyset, V \in C$ and C is closed under arbitrary intersections and nested unions.

Although convex sets have been introduced in different settings, the most useful definitions are based on the notion of *betweenness*. For example, given a metric space (X,d), a set A is *d-convex* if for every two points $x, y \in A$, the segment $[x,y]$ connecting x and y lies completely in A, where the segment $[x,y]$ stands for the set of points which are *between x and y*, i.e., $[x,y] = \{z \in X : d(x,z) + d(z,y) = d(x,y)\}$.

Given a finite, simple, connected graph $G = (V,E)$, it can be treated as a metric space by means of their shortest paths (also known as *geodesics*), as follows. For any pair of vertices $u, v \in V$, the distance between u and v is the length of any geodesic joining x and y. The convexity induced by this metric is called the *geodesic convexity* of G.

The main purpose of this lecture notes is to survey the main ideas, concepts, and results related to *geodesic convexity* in graphs. A glance at the table of contents shows that the book is organized in seven chapters; it contains a complete bibliography on the subject, a short glossary of definitions directly related to this topic, an index of terms, and finally, an index of symbols. Furthermore, unless otherwise expressly stated, all terms, invariants, and results pertain to the *geodesic convexity*.

In order to make the book as self-contained as possible, the monograph begins with an introductory chapter organized in four short sections: *Graph Theory, Metric Graph Theory, Convexity Spaces, and Graph Convexities*. Each of them ends up with a list of reference citations including books, surveys, and the most significant papers related to the topic.

The rest of the book is organized in six chapters. The first of them, under the title *Invariants*, contains the most basic and significant terms, concepts, results, and techniques, and hence it constitutes, roughly speaking, the core of the monograph. Except in a few cases, mainly due to space reasons, each nonobvious result is followed by the corresponding detailed proof, written in concise and schematic style, always starting from the original proof of the cited paper.

After carefully analyzing each of the more than 200 papers that have approached, directly or indirectly, the study of the geodesic convexity in graphs, most of them being published in the last two decades, and excluding those results included in Chap. 2, the rest of them have been grouped in five short chapters: *Graph Operations*, *Boundary Sets*, *Steiner Trees*, *Oriented Graphs*, and *Computational Complexity*. Each of them has been designed in the same way as Chap. 2, that is to say, including, in almost every case, result plus proof.

This monograph contains about 10 tables, 40 figures, 100 proofs, 25 sketches of proofs, 20 conjectures (or open problems), and 182 cited references.

To finalize, I wish to express my acknowledgement, for one or other reason, to the following persons: Danilo Artigas, Mónica Blanco, José Cáceres, Manoj Changat, Victor Chepoi, Carmen Hernando, Ferdinand P. Jamil, Changhong Lu, Janaina Minelli, Mercè Mora, Norbert Polat, Mari Luz Puertas, Seiichi Yamaguchi, Pei-Lan Yen, and Ping Zhang.

Barcelona, Spain   Ignacio M. Pelayo

Contents

Chapter 1
Introduction

1.1 Graph Theory

By a *graph* we mean[1] an ordered pair $G = (V, E)$, where $V = V(G)$ is a finite nonempty set of objects called *vertices* and $E = E(G)$ is a set of unordered pairs of distinct vertices, i.e., two-element subsets of V called *edges*. The cardinality $|V|$ of V and the cardinality $|E|$ of E are called the *order* and *size* of G, respectively. If $e = uv \in E(G)$, then any of the following statements may be used: u and v are adjacent, e joins u and v, e is incident with u (and with v), and u and v are neighbors.

The *(open) neighborhood* $N(v)$ of a vertex $v \in V(G)$ is the set of neighbors of v in G, i.e., $N(u) = \{v \in V(G) : uv \in E(G)\}$. The set $N[v] = N(v) \cup \{v\}$ is the *closed neighborhood* of v. The *degree* $\deg(v)$ of a vertex $v \in V(G)$ is the number of neighbors of v, i.e., $\deg(v) = |N(v)|$. The *maximum degree* of G, denoted by $\Delta(G)$, is the largest degree among all vertices of G. The *minimum degree* of G, denoted by $\delta(G)$, is the smallest degree among all vertices of G. Vertices of degree 0 and 1 are called *isolated vertices* and *end-vertices*, respectively. A graph G is called *regular* of degree r, or simply r-regular, if $\delta(G) = \Delta(G)$.

Let $[n]$ denote the set $\{1, \ldots, n\}$, for any positive integer n. The *complete graph* K_n is the graph of order n in which any two distinct vertices are adjacent. The *path* P_n of order n is the graph such that if $V(P_n) = [n]$, then $ij \in E(P_n)$ if and only if $|i - j| = 1$. The *cycle* C_n of order $n \geq 3$ is the two-regular graph obtained from P_n by joining vertices 1 and n.

By a *subgraph* H of a graph G we mean a graph H such that $V(H) \subseteq V(G)$ and $E(H) \subseteq E(G)$, in which case we write $H \subseteq G$. If $V(H) = V(G)$, then H is a *spanning subgraph* of G. An *induced subgraph* H of a graph G is any subgraph satisfying the following property: for every pair u, v of vertices of H, if they are adjacent in G, then they are also adjacent in H. If H is an induced subgraph of a graph G and $S = V(H)$, then we say that H is the subgraph induced by S in G and we write $H = G[S]$.

---

[1] Unless otherwise stated.

I.M. Pelayo, *Geodesic Convexity in Graphs*, SpringerBriefs in Mathematics, DOI 10.1007/978-1-4614-8699-2_1, © Ignacio M. Pelayo 2013

Given $S \subset V(G)$, the subgraph $G - S$, obtained by deleting all vertices in S from G, is the induced subgraph $G[V(G) - S]$. In particular, given $v \in V(G)$, the subgraph $G - v$, obtained by deleting v from G, is the induced subgraph $G[V(G) - v]$.

Two graphs G and H are *isomorphic*, in symbols $G \cong H$, if there exists a bijection ϕ from $V(G)$ onto $V(H)$ such that $uv \in E(G)$ if and only if $\phi(u)\phi(v) \in E(H)$. The *complement* $\overline{G}$ of a graph G is the graph with vertex set $V(\overline{G}) = V(G)$ such that two vertices are adjacent if and only they are not adjacent in G.

Let G_1, G_2 two graphs having disjoint vertex sets. The (disjoint) *union* $G = G_1 + G_2$ is the graph such that $V(G) = V(G_1) \cup V(G_2)$ and $E(G) = E(G_1) \cup E(G_2)$. The *join*[2] $G = G_1 \vee G_2$ is the graph such that $V(G) = V(G_1) \cup V(G_2)$ and $E(G) = E(G_1) \cup E(G_2) \cup \{uv : u \in V(G_1), v \in V(G_2)\}$. Notice that $G_1 \vee G_2 = \overline{\overline{G_1} + \overline{G_2}}$.

A set U of vertices of a graph G is called *independent* if no two vertices in U are adjacent, i.e., if $G[U] \cong \overline{K}_{|U|}$. A graph of order n is called *bipartite* if its vertex set can be partitioned into two independent sets U and W. The *complete bipartite graph* $K_{h,k}$ is the bipartite graph of order $n = h + k$ such that $V(K_{h,k}) = U \cup W$, $G[U] \cong \overline{K}_h$, $G[W] \cong \overline{K}_h$, and being every vertex in U adjacent to every vertex in W, i.e., $K_{h,k} = \overline{K}_h \vee \overline{K}_k$.

By a *k-path*, a *k-cycle*, and a *k-clique* of a graph G we mean a subgraph of order k in G which is isomorphic to the path P_k, to the cycle C_k, and to the complete graph K_k, respectively. For any two vertices $u, v \in V(G)$, a $u - v$ path is a path of G whose end-vertices are u and v, in which case we say that vertices u and v are connected in G. The *length* of a $u - v$ path P is the size of P, i.e., the number of edges it contains.

A *connected* graph[3] is a graph in which every two vertices are connected. A connected subgraph H of a graph G is a *component* of G if H is not a proper subgraph of any connected subgraph of G. A set S of vertices of a connected graph G is called a *cutset* of G if the graph $G - S$ is not connected. In particular, a vertex $v \in V(G)$ is a *cut-vertex* of G if $G - v$ is disconnected. The *connectivity* $\kappa(G)$ of a connected non-complete graph G is the minimum cardinality of a cutset of G. A bipartite graph is said to be a *tree* if it is connected and in addition it contains no subgraph isomorphic to a cycle, i.e., a graph such that every two vertices u and v are connected by exactly one $u - v$ path.

For additional details and information on basic graph theory we refer the reader to [14, 71, 108, 141].

1.2 Metric Graph Theory

For any two vertices $u, v \in V(G)$ of a connected graph G, a $u - v$ *geodesic* is a shortest $u - v$ path, i.e., a $u - v$ path of minimum order. The distance $d_G(u, v)$

---

[2]We follow the notation of [71]. Other authors denote the union and the join of a pair of graphs G and H by $G \cup H$ and $G + H$, respectively.

[3]Unless specified otherwise, every graph is supposed to be connected.

between vertices u and v is the length of a $u - v$ geodesic. When the graph G is clear we simply write $d(u,v)$. The distance function $d_G : G \times G \to \mathbb{N}$ associated to a connected graph G satisfies, for every $u,v,w \in V(G)$, the following properties:

1. $d(u,v) \geq 0$, equality holding if and only if $u = v$.
2. $d(u,v) = d(v,u)$.
3. $d(u,v) \leq d(u,w) + d(w,v)$.

Hence, for every connected graph G, d_G is a metric on $V(G)$ and the pair $(V(G), d_G)$ is a metric space. Associated with this graph metric are special types of subgraphs and vertices that have played an important role in the literature. We devote the remainder of this section to such topics.

For every induced subgraph H of a graph G and every pair of vertices $u,v \in V(H)$, the inequality $d_G(u,v) \leq d_H(u,v)$ holds. If $d_G(u,v) = d_H(u,v)$ for any $u,v \in V(H)$, then H is said to be an *isometric subgraph* of G. More generally, an injective mapping f from the vertex set $V(H)$ of a graph H to the vertex set $V(G)$ of a graph G is called an *isometric embedding* if $d_G(f(u),f(v)) = d_H(u,v)$ for any $u,v \in V(H)$.

The *eccentricity* $e(v)$ of a vertex $v \in V(G)$ is the maximum distance between v and any other vertex of G, i.e., $e(v) = \max\{d(v,u) : u \in V(G)\}$. The *radius* rad$(G)$ of G is the smallest eccentricity among the vertices of G, i.e., rad$(G) = \min\{e(v) : v \in V(G)\}$. A *central* vertex is any vertex v such that $e(v) = $ rad(G). The *center* Cen(G) is the subgraph induced by the set of central vertices of G.

The *diameter* diam(G) of G is the maximum distance between two vertices of G, i.e., diam$(G) = \max\{e(v) : v \in V(G)\}$. Observe that rad$(G) \leq$ diam$(G) \leq 2$rad(G). Any vertex v such that $e(v) = $ diam(G) is called *peripheral*, whereas any two vertices at a distance equal to the diameter of the graph are said to be *antipodal*. Notice that a vertex v is peripheral if and only if the positive integer-valued function $e : V(G) \mapsto \mathbb{N}$ has a global maximum at v. The *periphery* Per(G) is the subgraph induced by the set of peripheral vertices of G. A vertex $v \in V(G)$ is called *locally peripheral*[4] if e has a local maximum at v, i.e., if $e(v) \geq e(u)$ for every $u \in N(v)$. The *contour* Ct(G) is the subgraph induced by the set of locally peripheral vertices of G [23].

A vertex $v \in V(G)$ is called *eccentric* if, for some vertex $u \in V(G)$, the integer-valued function $d(u,-)$ has a global maximum at v, i.e., if there exists another vertex $u \in V(G)$ such that no vertex in the entire graph is further away from u than v. The *eccentric set* Ecc(u) of a vertex u is the set of all its eccentric vertices. The *eccentric subgraph* Ecc(G) is the subgraph induced by the set of eccentric vertices of G, i.e., Ecc$(G) = G[\cup_{u \in G}$Ecc$(u)]$. A vertex $v \in V(G)$ is said to be *locally eccentric*[5] if, for some vertex $u \in V(G)$, the function $d(u,-)$ has a local maximum at v, i.e., if there exists another vertex $u \in V(G)$ such that no neighbor of v is further away from u than v. By $\partial(u)$ we denote the set of all locally eccentric vertices of a given vertex u, and the vertices of $\partial(u)$ are said to be the *maximally distant vertices* from u [151].

---

[4] Also known as *contour vertex*.

[5] Also known as *boundary vertex*.

Fig. 1.1 Venn diagram of
$\partial(G)$, Ecc(G), Ct(G),
Per(G), and Ext(G)

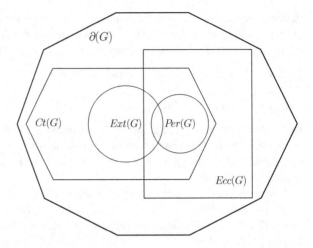

The *boundary* $\partial(G)$ is the subgraph induced by the set of locally eccentric vertices
of G, i.e., $\partial(G) = \bigcup_{u \in V(G)} \partial(u)$ [66].

A vertex v of a graph G is called *simplicial* if its neighborhood $N(v)$ induces
a clique, i.e., if $G[N(v)] \cong K_{\deg(v)}$. The *extreme subgraph* Ext$(G)^6$ is the subgraph
induced by the set of simplicial vertices of G. Notice that (see Fig. 1.1):

1. Every simplicial vertex is locally peripheral, i.e., Ext$(G) \subseteq$ Ct(G).
2. Every peripheral vertex is locally peripheral, i.e., Per$(G) \subseteq$ Ct(G).
3. Every peripheral vertex is eccentric, i.e., Per$(G) \subseteq$ Ecc(G).
4. Every eccentric vertex is locally eccentric, i.e., Ecc$(G) \cup \partial(G)$.
5. Every locally peripheral vertex is locally eccentric, i.e., Ct$(G) \subseteq \partial(G)$.

Proposition 1.1 ([24]). *Let $G = (V, E)$ be a connected graph.*

1. *If $|\text{Per}(G)| = |\text{Ct}(G)| = 2$, then either $|\partial(G)| = 2$ or $|\partial(G)| \geq 4$.*
2. *If $|\text{Ecc}(G)| = |\text{Per}(G)| + 1$, then $|\partial(G)| > |\text{Ecc}(G)|$.*
3. *If $|\text{Ecc}(G)| > |\text{Per}(G)|$, then $|\partial(G)| \geq |\text{Per}(G)| + 2$.*

Proof.

1. Suppose that $|\partial(G)| = 3$, that is, Per$(G) = $ Ct$(G) = \{a, b\}$ and $\partial(G) = \{a, b, c\}$.
   Check that the vertex c lies on some $a - b$ geodesic P. Let W be the set of all
   vertices of which c is a boundary vertex. Notice that $W \cap V(P) = \emptyset$. Take a vertex
   $y \in W$ satisfying $d(y, c) = \max_{x \in W} d(x, c) = h$.

   Certainly, $y \notin \partial(G)$, since $y \notin V(P)$ and $|\partial(G)| = 3$. In particular, y is not
   a boundary vertex of the vertex c, i.e., there exists a neighbor z of y such that
   $d(c, z) = h + 1$ (see Fig. 1.2). Notice that $z \notin V(P)$ since c is a boundary vertex

---

^{6}Ext(G), Per(G), Ecc(G), Ct(G), and $\partial(G)$ also denote the corresponding set of vertices.

Fig. 1.2 $d(z,c) = d(z,y)+d(y,z) = 1+h$

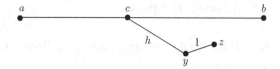

of y. If w is an arbitrary neighbor of the vertex c, then $d(z,w) \leq d(z,y)+d(y,w) \leq 1+h = d(z,c)$. Hence, we have proved that z is a boundary vertex of c, which is a contradiction.

2. Let $x \in \mathrm{Ecc}(G) \setminus \mathrm{Per}(G)$. Take the set $W = \{y \in V : d(y,x) = e(y)\}$. Notice that $W \cap \mathrm{Per}(G) = \emptyset$, since $x \notin \mathrm{Per}(G)$. Observe also that $W \cap \mathrm{Ecc}(G) = \emptyset$, since $\mathrm{Ecc}(G) = \mathrm{Per}(G) \cup \{x\}$. Consider a vertex $z \in W$ such that $e(z) = \max_{y \in W} e(y)$.

   In order to prove that z is a boundary vertex of x, let us suppose, on the contrary, that there exists a vertex $w \in N(z)$ such that $d(w,x) = d(z,x)+1$. It means both that $e(w) = e(z)+1$ and $w \in W$, which is a contradiction. Hence, $z \in \partial(G)$, as desired.

3. This result is a corollary of the previous one since $\mathrm{Per}(G) \subset \mathrm{Ecc}(G) \subseteq \partial(G)$. □

Corollary 1.1 ([24]). *Let G be a nontrivial connected graph such that $|\mathrm{Per}(G)| = a$, $|\mathrm{Ct}(G)| = b$, $|\mathrm{Ecc}(G)| = c$, and $|\partial(G)| = d$. Then,*

$$\begin{cases} 2 \leq a \leq b \leq d, \\ 2 \leq a \leq c \leq d, \\ (a,b,c,d) \neq (2,2,2,3), \\ (a,b,c,d) \neq (a,b,a+1,a+1). \end{cases}$$

Theorem 1.1 ([24, 25]). *Let $(a,b,c,d) \in \mathbb{N}^4$ integers satisfying the conditions of Corollary 1.1. Then, there exists a connected graph $G = (V,E)$ satisfying:*

$$|\mathrm{Per}(G)| = a, \quad |\mathrm{Ct}(G)| = b, \quad |\mathrm{Ecc}(G)| = c, \quad |\partial(G)| = d.$$

Proposition 1.2 ([117, 118]). *Let G be a nontrivial connected graph such that $|\mathrm{Ext}(G)| = a$, $|\mathrm{Per}(G)| = b$, $|\mathrm{Ct}(G)| = c$, and $|\partial(G)| = d$. Then,*

$$\begin{cases} 0 \leq a \leq c \leq d, \\ 2 \leq b \leq c \leq d, \\ (a,b,c,d) \neq (a,2,2,3), \\ (a,b,c,d) \neq (2,2,3,3), \\ (a,b,c,d) \neq (a,3,3,3). \end{cases}$$

Theorem 1.2 ([117,118]). *Let $(a,b,c,d) \in \mathbb{N}^4$ integers satisfying the conditions of Proposition 1.2. If $a \geq 2$, then there exists a connected graph $G = (V,E)$ satisfying:*

$$|\mathrm{Ext}(G)| = a, \quad |\mathrm{Per}(G)| = b, \quad |\mathrm{Ct}(G)| = c, \quad |\partial(G)| = d.$$

For additional details and information on metric graph theory we refer the reader to [10, 11, 21, 23–25, 56, 66, 114].

1.3   Convexity Spaces

A *convexity* $\mathscr{C}$ on a nonempty set V is a collection of subsets of V such that:

(C1) $\emptyset, V \in \mathscr{C}$.
(C2)   Arbitrary intersections of convex sets are convex.
(C3)   Every nested union of convex sets is convex.

A *convexity space*[7] is an ordered pair $(V, \mathscr{C})$, where V is a nonempty set and $\mathscr{C}$ is a convexity on V. The members of $\mathscr{C}$ are called *convex sets*.

Given a set $S \subset V$, the smallest convex set $[S]_{\mathscr{C}}$ containing S is called the *convex hull* of S. If $[S]_{\mathscr{C}} = V$, then S is said to be $\mathscr{C}$-*hull set* of V. A fundamental property of the convex hull operator $[\ ]_{\mathscr{C}}$ associated with a convexity $\mathscr{C}$ is to be a *finitary*[8] *closure operator*.

Theorem 1.3 ([89]). *Let $\mathscr{C}$ be a collection of subsets of a nonempty set V satisfying axioms (C1) and (C2). Then,*

- $[\ ]_{\mathscr{C}}$ *is a closure operator, i.e., for all sets $A, B \subseteq V$,*

    - $A \subseteq [A]_{\mathscr{C}}$.
    - *If $A \subseteq B$, then $[A]_{\mathscr{C}} \subseteq [B]_{\mathscr{C}}$.*
    - $[[A]_{\mathscr{C}}]_{\mathscr{C}} = [A]_{\mathscr{C}}$.

- *Axiom (C3) holds if and only if $\mathscr{C}$ satisfies the finitary axiom:*

    (C3') If $x \in [A]_{\mathscr{C}}$, then $x \in [F]_{\mathscr{C}}$ for some finite A-subset F.

Basic Examples

- *Trivial convexity*: $V \neq \emptyset$, $\mathscr{C} = \{\emptyset, V\}$.
- *Free convexity*: $V \neq \emptyset$, $\mathscr{C} = 2^V$.
- $V = \mathbb{N}, k \geq 1, \mathscr{C} = \{A \subset V : |A| \leq k\} \cup \{V\}$.
- *Standard convexity* in a real vector space V:

$$C \subseteq V \text{ is convex} \Leftrightarrow \forall x, y \in C, \forall t \in [0,1]: t \cdot x + (1-t) \cdot y \in C.$$

- *Order convexity* in a poset $(V, \leq)$:

$$C \subseteq V \text{ is order convex} \Leftrightarrow \forall x, y \in C: x \leq z \leq y \Rightarrow z \in C.$$

- *Metric convexity* in a metric space (V, d):

$$C \subseteq V \text{ is convex} \Leftrightarrow \forall x, y \in C, \{z \in V : d(x,z) + d(z,y) = d(x,y)\} \subseteq C.$$

---

[7] Also known as *aligned space* and *alignment*.
[8] Also known as *domain finite*.

- Some metric convexities:

  - Euclidean convexity in $\mathbb{R}^n$:

    metric convexity in $(\mathbb{R}^n, d_2)$, where $d_2(x,y) = (\sum_{i=1}^n (x_i - y_i)^2)^{\frac{1}{2}}$.

  - Manhattan convexity in $\mathbb{R}^n$:

    metric convexity in $(\mathbb{R}^n, d_1)$, where $d_1(x,y) = \sum_{i=1}^n |x_i - y_i|$.

  - Chebyshev convexity in $\mathbb{R}^n$:

    metric convexity in $(\mathbb{R}^n, d_\infty)$, where $d_\infty(x,y) = \max_{1 \le i \le n} |x_i - y_i|$.

Given a convexity space $(V, \mathscr{C})$ and a convex set $W \subseteq V$, a vertex $v \in W$ is called an *extreme point* of W if the set $W \setminus \{v\}$ is also convex. A *convex geometry* (or an *antimatroid*) is a convexity space satisfying the so-called *Krein-Milman* property: *Every convex set is the convex hull of its extreme points* [100, 127, 134].

A convexity space $(V, \mathscr{C})$ satisfies the *anti-exchange* property if for any convex set W and two distinct points $x, y \notin W$, then at most one of $y \in [W \cup \{x\}]_{\mathscr{C}}$ and $x \in [W \cup \{y\}]_{\mathscr{C}}$ holds true.

Theorem 1.4 ([127, 135]). *For every convexity space $(V, \mathscr{C})$, the following three statements are equivalent:*

- $(V, \mathscr{C})$ *has the anti-exchange property.*
- *If $W \in \mathscr{C}$ and $W \neq V$, then $W \cup \{y\} \in \mathscr{C}$ for some $y \in V - W$.*
- $(V, \mathscr{C})$ *is a convex geometry.*
- *If $W \in \mathscr{C}$ and $x \notin W$, then x is an extreme point of $[W \cup \{x\}]_{\mathscr{C}}$.*

For additional details and information on abstract and general convexity theory we refer the reader to [32, 75, 90, 94, 106, 127, 133–135, 175].

1.4 Graph Convexities

A *graph convexity space* is an ordered pair $(G, \mathscr{C})$, formed by a connected graph $G = (V, E)$ and a convexity $\mathscr{C}$ on V such that $(V, \mathscr{C})$ is a convexity space satisfying the additional axiom [89] :

(C4)   Every member of $\mathscr{C}$ induces a connected subgraph of G.

Let ϕ be a property to be checked on the set of paths in a graph G. A ϕ-path is a path having property ϕ. A path property is called *feasible* if, for every pair of distinct vertices $u, v \in V(G)$, there is a ϕ-path joining u and v. The most natural graph convexities are the so-called path convexities (a type of interval convexity [144]) defined by a family $\mathscr{P}_\phi$ of paths, being ϕ a feasible path property.

Let ϕ be a feasible path property and let $\mathcal{P}_\phi$ be the correspondent family of ϕ-paths on a given graph G. If S is a set of vertices of G, then the ϕ-*closure* $I_\phi[S]$ consists of S together with all vertices lying on some ϕ-path joining two vertices of S. A set $S \subseteq V(G)$ is called ϕ-*convex* if $I_\phi[S] = S$.

Main Path Convexities

- *Geodesic convexity*: Also known as *metric* convexity and *d*-convexity. It is the path convexity obtained when $\mathcal{P}_\phi$ is the set of all geodesics, i.e., when ϕ is the property of being a shortest path [112, 113, 148].
- *Detour convexity*: Also known as *longest path* convexity. It is the path convexity obtained when ϕ is the property of being a longest path [46, 68, 70]. Notice that this convexity induces, as in the previous case, a metric on the graph.
- *Monophonic convexity*: Also known as *induced path* covexity and *minimal path* convexity. A *chord* of a path P in a graph G is any edge joining a pair of nonadjacent vertices of P. This convexity is the path convexity obtained when ϕ is the property of being a *chordless path*, i.e., an induced path [41, 89, 100].
- *All-paths convexity*: It is the path convexity obtained when $\mathcal{P}_\phi$ is the set of all paths. It is the coarsest path convexity [42, 107, 165].
- *Triangle-path convexity*: It is the path convexity obtained when $\mathcal{P}_\phi$ is the set of all paths allowing just *short chords*, i.e., those chords joining vertices at distance 2 apart in each of these paths [40, 43, 45].
- *Total convexity*: Also known as *triangle-free path* convexity. It is the path graph convexity obtained when $\mathcal{P}_\phi$ is the set of all paths allowing just *long chords*, i.e., those chords joining vertices at distance at least 3 apart in each of these paths [76].

Some authors have introduced and studied a number of path convexities for which axiom (C4) does not hold, since the selected path property is not feasible. Some of them are the following:

- m^3-*convexity*: It is the path convexity obtained when $\mathcal{P}_\phi$ is the set of all induced paths of length at least 3 [28, 88].
- g^3-*convexity*: It is the path convexity obtained when $\mathcal{P}_\phi$ is the set of all shortest paths of length at least 3 [150].
- P_3-*convexity*: Also known as *two-path convexity*. It is the path graph convexity obtained when $\mathcal{P}_\phi$ is the set of all paths of length 2 [37, 85, 153].
- g_k-*convexity*: It is the path convexity obtained when $\mathcal{P}_\phi$ is the set of all shortest paths of length at most k [101].

For additional details and information on graph convexity theory we refer the reader to [43, 44, 89, 100, 142–144, 150, 156, 169].

Chapter 2
Invariants

2.1 Geodetic Closure and Convex Hull

For two vertices u and v of a graph G, a vertex $x \in V(G)$ is said to be *geodominated* by the pair $\{u,v\}$ if x lies on some $u - v$ geodesic in G. The *geodetic interval* $I_G[u,v]$ consists of u,v together with all vertices geodominated by the pair $\{u,v\}$. If S is a set of vertices of G, then the *geodetic closure* $I_G[S]$ is the union of all sets $I[u,v]$ for $u,v \in S$, i.e., it consists of S together with all vertices lying on some geodesic joining two vertices of S. When the graph G is clear from the context, $I_G[u,v]$ and $I_G[S]$ are usually replaced by $I[u,v]$ and $I[S]$, respectively.

A set S of vertices is called *geodesically convex*, *g-convex*, or simply *convex*, if $I[S] = S$, i.e., if for every pair $u,v \in S$, the interval $I[u,v] \subseteq S$. In any graph, the empty set, the whole vertex set, every singleton, and every two-path are convex. On the other hand, if $I[S] = V(G)$, then S is said to be a *geodetic set* (also known as *geodominating set*).

Let W be a set of vertices of a graph G. We define $I^k[W]$ recursively as follows: $I^0[W] = W$, $I^1[W] = I[W]$, and $I^k[W] = I[I^{k-1}[W]]$ for $k > 1$. Since the vertex u forms the only $u - u$ geodesic, we have $I[u,u] = u$, and hence $W \subseteq I[W]$. Observe that this implies that W is convex if and only if $I[W] = W$. The *geodetic iteration number* $\mathrm{gin}(W)$ of W is the smallest positive integer n such that $I^n[W] = I^{n+1}[W]$. Note that $I[W]$ is convex if and only if $\mathrm{gin}(W) = 1$. The *geodesic iteration number* of G, denoted by $\mathrm{gin}(G)$, is defined as $\mathrm{gin}(G) = \max\{\mathrm{gin}(W) : W \subseteq V(G)\}$.

The *convex hull* $[S]$ of a set S of vertices is the smallest convex set containing S. It is also called the *geodesic convex hull* or simply the *g-convex hull* of S and denoted by $[S]_g$. As an immediate consequence of this definition, the following properties hold.

Proposition 2.1. *Let S be a non-convex set of a graph G. Then, $S \subset I[S] \subseteq [S] \subseteq V(G)$. Moreover, the following statements are equivalent:*

- *$[S]$ is the smallest convex set containing S.*
- *$[S]$ is the intersection of all convex sets containing S.*

I.M. Pelayo, *Geodesic Convexity in Graphs*, SpringerBriefs in Mathematics, DOI 10.1007/978-1-4614-8699-2_2, © Ignacio M. Pelayo 2013

- $[S] = I^k[S]$, for every positive integer $k \geq \mathrm{gin}(S)$.

If a set S satisfies $[S] = V(G)$, then it is called a *hull set* of G. Clearly, every geodetic set is a hull set, but the converse is not necessarily true. For example, if $U = \{x,y\}$ and $W = \{a,b,c\}$ are the partite sets of $K_{2,3}$, then the set $\{a,b\}$ is a hull set as $I^2[a,b] = V$, but it is not geodetic since $I[a,b] = \{a,b,x,y\}$.

2.2 Geodetic and Hull Numbers

The *geodetic number* of a graph G, denoted by $g(G)$, is the minimum cardinality of a geodetic set of $V(G)$ [113]. The *hull number* of a graph G, denoted by $h(G)$, is the minimum cardinality of a hull set of $V(G)$ [98].

If $G = \sum_{i=1}^{p} G_i$ is a non-connected graph with p components, then $g(G) = \sum_{i=1}^{p} g(G_i)$ and $h(G) = \sum_{i=1}^{p} h(G_i)$. Hence, the study of both parameters can be restricted to connected graphs.

Theorem 2.1 ([63]). *If G is a nontrivial graph of order n and diameter d, then $2 \leq h(G) \leq g(G) \leq n - d + 1$.*

Proof. Every nontrivial graph satisfies $2 \leq h(G) \leq g(G)$, since every geodetic set is a hull set and the unique graph for which $h(G) = 1$ is $G \cong K_1$. Let u, v be vertices of G for which $d(u,v) = d$ and ρ a $u - v$ geodesic. If $V(\rho) = \{u, v_1, \ldots, v_{d-1}, v\}$ and $S = V(G) \setminus \{v_1, \ldots, v_{d-1}\}$, then $I[S] = V(G)$ and, consequently, $g(G) \leq |S| = n - d + 1$. □

A corollary of this result is that the only connected graph of order and geodetic number n is the complete graph K_n. In Table 2.1, both the hull number and the geodetic number of some basic graphs are displayed.

Theorem 2.2 ([59, 63]). *If n, d, and k are integers such that $2 \leq d < n$, $2 \leq k < n$, and $n - d - k + 1 \geq 0$, then there exists a graph G of order n, diameter d, and $h(G) = g(G) = k$.*

Proof. Let P_{d+1} a path of order $d + 1$ such that $V(P_{d+1}) = \{u_0, u_1, \ldots, u_d\}$. We first add $k - 2$ new vertices $v_1, v_2, \ldots, v_{k-2}$ to P_{d+1}, and join each to u_1, producing a tree T. Then, we add $n - d - k + 1$ new vertices $w_1, w_2, \ldots, w_{n-d-k+1}$ and join each to both u_0 and u_2, thereby producing the graph G or order n and diameter d displayed

Table 2.1 Geodetic and hull numbers of some basic graph families

G[a]	P_n	C_{2l}	C_{2l+1}	T_n	K_n	$K_{p,q}$[b]	W_n[c]	Q_n
$h(G)$	2	2	3	$|\mathrm{Ext}(T_n)|$	n	2	$\lfloor \frac{n}{2} \rfloor$	2
$g(G)$	2	2	3	$|\mathrm{Ext}(T_n)|$	n	$\min\{4,p\}$	$\lfloor \frac{n}{2} \rfloor$	2

[a] $G \not\cong K_1$
[b] $2 \leq p \leq q$
[c] $n \geq 5$

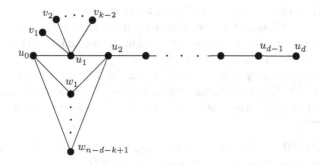

Fig. 2.1 $h(G) = g(G) = k$

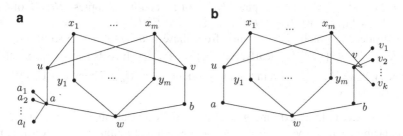

Fig. 2.2 (a) Graph $J_{7,m,l}$, (b) graph $J_{7,m}^k$

in Fig. 2.1. Finally, it is a routine exercise to check that $S = \{u_0, u_d, v_1, \ldots, v_{k-2}\}$ is both a minimum geodetic set and a minimum hull set of G. $\square$

Theorem 2.3 ([59, 116]). *For every pair* α, β *of integers with* $2 \leq \alpha \leq \beta$, *there exists a graph G such that* $h(G) = \alpha$ *and* $g(G) = \beta$.

Proof. For $\alpha = \beta$, K_α has the desired properties. Suppose that $\alpha < \beta$ and consider the graph $J_{7,m,l}$ displayed in Fig. 2.2a. Observe that $\{v\} \cup \{a_j\}_{j=1}^l$ $\left(\text{resp. } \{v\} \cup \{a_j\}_{j=1}^l \cup \{y_i\}_{i=1}^m\right)$ is a minimum hull (resp. geodetic) set. Thus, taking $m = \beta - \alpha$ and $l = \alpha - 1$, we have a graph G satisfying $h(G) = \alpha$, $g(G) = \beta$. $\square$

Proposition 2.2 ([98]). *A vertex x is an extreme vertex with respect to the geodesic convexity in a graph G if and only if it is simplicial.*

Proof. Let x be an extreme vertex. If $\deg_G(x) = 1$, then x is trivially simplicial. Suppose that $\deg_G(x) \geq 2$ and take a pair of vertices $u, v \in N(x)$. Since $V - x$ is convex, every vertex lying on some $u - v$ geodesic must belong to $V - x$, which is only true if u and v are adjacent.

Conversely, assume that x is a simplicial vertex of G and take a pair of vertices $u, v \in V(G) - x$. Let ρ be a $u - v$ geodesic containing vertex x and take the neighbors $\{a, b\}$ of x belonging to $V(\rho)$. Notice that $d_G(a, b) = 2$, a contradiction since $N(x)$ induces a clique. $\square$

Every hull set of a graph $G = (V, E)$, and hence also every geodetic set, contains all vertices of $\text{Ext}(G)$ since for every extreme vertex x and for every set S such that $S \subseteq V - x$, its convex hull $[S]$ is also contained in $V - x$ (see [98]). In other words, every graph G satisfies $|\text{Ext}(G)| \leq h(G) \leq g(G)$.

An *extreme geodesic graph* is a graph G such that $|\text{Ext}(G)| = g(G)$. In these graphs, the set $\text{Ext}(G)$ of simplicial vertices is both its unique minimum geodetic and hull set. Complete graphs and trees provide two basic examples of extreme geodesic graphs.

Theorem 2.4 ([54]). *For every pair a, b of integers with $(a, b) \neq (0, 1)$ and $0 \leq a \leq b$, there exists a graph G such that $|\text{Ext}(G)| = a$ and $g(G) = b$.*

Proof. If $1 \leq a = b$, then K_a satisfies $|\text{Ext}(K_a)| = g(K_a) = a$. Thus we assume that $0 \leq a < b$ and $2 \leq b$. Take a copies $\{F_i\}_{i=1}^{a}$ of K_2 and $b - a$ copies $\{H_i\}_{i=1}^{b-a}$ of C_4. Let $V(F_i) = \{x_i, y_i\}$, $V(H_j) = \{u_j, v_j, w_j, z_j\}$, and $E(H_j) = \{u_j v_j, v_j w_j, w_j z_j, z_j u_j\}$ and consider the graph $G_{a,b}$ obtained from these two families by identifying the vertices $\{x_i\}_{i=1}^{a} \cup \{u_j\}_{j=1}^{b-a}$. Clearly, $\text{Ext}(G_{a,b}) = \{y_i\}_{i=1}^{a}$. It is also easy to check that the set $\{y_i\}_{i=1}^{a} \cup \{w_j\}_{j=1}^{b-a}$ is a minimum geodetic set of $G_{a,b}$. Thus, $|\text{Ext}(G_{a,b})| = a$ and $g(G_{a,b}) = b$. □

Another example of extreme geodesic graph is that obtained from a star $K_{1,k}$ by replacing each end-vertex by a complete graph and joining all new vertices to the cut-vertex v of the star. Observe that the extreme subgraph of any graph G of order n belonging to this family contains all vertices of G but v, which means that $h(G) = g(G) = n - 1$. Moreover, as shown next, these are the only graphs verifying this double equality.

Theorem 2.5 ([22, 59, 98]). *Let G be a connected graph of order $n \geq 3$. Then, the following three statements are equivalent:*

1. $h(G) = n - 1$.
2. $g(G) = n - 1$.
3. $G \cong K_1 \vee (K_{n_1} + \cdots + K_{n_r})$ where $r \geq 2$ and $n_1 + \cdots + n_r = n - 1$.

Proof. We shall show that $(1) \Rightarrow (2) \Rightarrow (3) \Rightarrow (1)$.
$(1) \Rightarrow (2)$ The only connected graph such that $g(G) = n$ is the complete graph K_n. Therefore, $g(G) = n - 1$, as $h(G) \leq g(G)$ and $h(K_n) = n$.
$(2) \Rightarrow (3)$ Let S be a minimum geodetic set of G, where $V(G) - S = v$. We claim that v is adjacent to every vertex in S. For every geodesic ρ joining any two nonadjacent vertices u, v of G, the set $[V(G) - V(\rho)] \cup \{u, v\}$ is a geodetic set of G. This means, first, that $\text{diam}(G) = 2$ and, second, that v is adjacent to every pair of nonadjacent vertices of S. Therefore, if u is a vertex of S nonadjacent to vertex v, then it must be adjacent to all other vertices of S. Since S is a geodetic set, v lies on some $s - t$ geodesic of length 2, where $s, t \in S$. Finally, since $us, ut \in E(G)$, it follows that $u \notin S$, a contradiction. Hence, as claimed, v is adjacent to every vertex in S. To show that $G \cong K_1 \vee (K_{n_1} + \cdots + K_{n_r})$, it suffices to observe that for any triple $x, y, z \in S$, if $xy, yz \in E(G)$, then $x, z \in E(G)$, since otherwise $d(x, z) = 2$, and hence $y = v$, a contradiction.

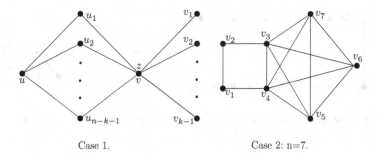

Case 1.   Case 2: n=7.

Fig. 2.3 Case 1: $2 \leq k \leq n-3$, Case 2: $2 \leq k = n-2$

$(3) \Rightarrow (1)$ All vertices of G but the cut-vertex of G are extreme vertices. Therefore, $h(G) = n-1$, since $|\text{Ext}(G)| \leq h(G)$ and the only connected graph such that $h(G) = n$ is K_n. $\qquad\square$

Corollary 2.1 ([54]). *Every nontrivial graph G of order n with geodetic number $g(G) = n-1$ is an extreme geodesic graph.*

Theorem 2.6 ([54]). *For every pair k, n of integers with $2 \leq k \leq n-2$ there exists a graph G of order n and geodetic number $g(G) = k$ that is not an extreme geodesic graph.*

Proof. We consider two cases.

Case 1: $2 \leq k \leq n-3$. Take $G_1 = K_{2,n-k-1}$ and $G_2 = K_{1,k-1}$. Consider the graph G obtained from G_1 and G_2 by identifying v and z, where $V(G_1) = \{u,v\} \cup \{u_i\}_{i=1}^{n-k-1}$ and $V(G_2) = \{z\} \cup \{v_i\}_{i=1}^{k-1}$ (see Fig. 2.3). Clearly, $\text{Ext}(G) = \{v_i\}_{i=1}^{k-1}$. It is also easy to check that the set $\{u\} \cup \{v_i\}_{i=1}^{k-1}$ is the unique minimum geodetic set of G. Thus, $|\text{Ext}(G)| = k-1 < k = g(G)$.

Case 2: $2 \leq k = n-2$. Take $G_1 = K_2$ and $G_2 = K_{n-2}$. Consider the graph G obtained from G_1 and G_2 by adding the edges $v_2 v_3$ and $v_1 v_4$, where $V(G_1) = \{v_1, v_2\}$ and $V(G_2) = \{v_i\}_{i=3}^{n}$ (see Fig. 2.3). Clearly, $\text{Ext}(G) = \{v_i\}_{i=4}^{n}$. It is also easy to check that the set $\{v_1, v_2\} \cup \{v_i\}_{i=4}^{n}$ is a minimum geodetic set of G. Thus, $|\text{Ext}(G)| = n-4 < n-2 = g(G)$. $\qquad\square$

Theorem 2.7 ([54]). *If r, d, and k are integers such that $2 \leq k$ and $r \leq d \leq 2r$, then there exists an extreme geodesic graph G such that $\text{rad}(G) = r$, $\text{diam}(G) = d$, and $h(G) = g(G) = k$.*

Proof. When $r = 1$, we let $G = K_k$ or $G = K_{1,k}$ according to whether $d = 1$ or $d = 2$, respectively. For $r \geq 2$, we construct an extreme geodesic graph G with the desired properties. Take the cycle C_{2r} and the path P_{d-r+1}, where $V(C_{2r}) = \{v_1, \dots, v_{2r}\}$ and $V(P_{d-r+1}) = \{u_0, u_1, \dots, u_{d-r}\}$. Let H be the graph obtained from C_{2r} and P_{d-r+1} by identifying $v_1 \in V(C_{2r})$ and $u_0 \in V(P_{d-r+1})$ and adding the edge $v_r v_{r+2}$. Finally, the graph G is then obtained by adding $k-2$ new vertices $w_1, w_2, \dots, w_{k-2}$ to H and joining all of them to the vertex u_{d-r+1} (see Fig. 2.4). Clearly, G is a graph of radius

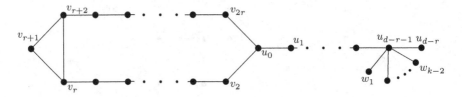

Fig. 2.4 Extreme geodesic graph G s.t. $\mathrm{rad}(G) = r$, $\mathrm{diam}(G) = d$, and $h(G) = g(G) = k$

r and diameter d. It is also easy to check that $\mathrm{Ext}(G) = \{v_{r+1}, w_1, \ldots, w_{k-2}, u_{d-r}\}$ is the sole geodetic set of G. Therefore, $g(G) = |\mathrm{Ext}(G)| = k$, as desired. $\square$

Theorem 2.8 ([54]). *If n, d, and k are integers such that $2 \leq d < n$, $2 \leq k < n$, and $n - d - k + 1 \geq 0$, then there exists an extreme geodesic graph G of order n, diameter d, and $g(G) = k$.*

Proof. Take the graph $H \cong \overline{K}_k \vee K_{n-d-k+2}$ where $V(\overline{K}_k) = \{v_1, \ldots, v_k\}$ and $V(K_{n-d-k+2}) = \{w_1, \ldots, w_{n-d-k+2}\}$. Consider the path P_{d-1}, where $V(P_{d-1}) = \{u_0, \ldots, u_{d-2}\}$. Let G be the graph obtained from H and P_{d-1} by identifying $v_1 \in V(H)$ and $u_0 \in V(P_{d-1})$. Certainly, G is a graph of order n and diameter d. It is also easy to check that the set $\mathrm{Ext}(G) = \{u_{d-2}, v_2, \ldots, v_k\}$ is the sole geodetic set of G. Therefore, $g(G) = |\mathrm{Ext}(G)| = k$, as desired. $\square$

Conjecture 2.1 ([54]). For any four integers a, b, d, n such that $a \leq b \leq n - 2$, $2 \leq b, d$, and $b + d - 1 \leq n$, there exists a graph G of order n and diameter d such that $|\mathrm{Ext}(G)| = a$ and $g(G) = b$.

The *geodetic ratio* and *extreme ratio* of a graph G of order n are defined respectively as $r_g(G) = \frac{g(G)}{n}$ and $r_{\mathrm{ext}}(G) = \frac{|\mathrm{Ext}(G)|}{n}$.

Theorem 2.9 ([54]). *For every pair s, t of rational numbers with $0 \leq s < t < \frac{1+s}{2} < 1$, there exists a connected graph G with $r_{\mathrm{ext}}(G) = s$ and $r_g(G) = t$.*

Proof. First, we assume that $s > 0$. Let $s = \frac{s_1}{s_2}$ and $t = \frac{t_1}{t_2}$, where s_1, s_2, t_1, t_2 are positive integers. Since $0 < s < t < \frac{1+s}{2} < 1$, it follows that $s_2 t_1 - s_1 t_2 > 0$ and $s_2 t_2 - 2 s_2 t_1 + s_1 t_2 > 0$. Take an even integer $k > 1$ and consider the integers $a = k s_1 t_2$, $k(s_2 t_1 - s_1 t_2)$, and $c = k(s_2 t_2 - 2 s_2 t_1 + s_1 t_2)$. Take $a - 1$ copies $\{F_i\}_{i=1}^a$ of K_2, b copies $\{H_i\}_{j=1}^b$ of $K_{2,3}$ and the path P_{c+1}. Consider the graph G obtained from all of these graphs by identifying the vertices $\{y_i\}_{i=1}^{a-1} \cup \{w_{j1}\}_{j=1}^b \cup \{v_{c+1}\}$, where $V(F_i) = \{x_i, y_i\}$, $V(H_j) = \{u_{j1}, u_{j2}\} \cup \{w_{j1}, w_{j2}, w_{j3}\}$, and $V(P_{c+1}) = \{v_h\}_{h=1}^{c+1}$. Clearly, the order of G is $n = a + 4b + c = k s_2 t_2$. It is also easy to check that $\mathrm{Ext}(G) = \{x_i\}_{i=1}^{a-1} \cup \{v_1\}$ and that the set $S = \{x_i\}_{i=1}^{a-1} \cup \{v_1\} \cup \{u_{j1}, u_{j2}\}_{j=1}^b$ is a minimum geodetic set of G. Hence, $|\mathrm{Ext}(G)| = a = k s_1 t_2$ and $g(G) = a + 2b = k s_2 t_1$, i.e., $r_{\mathrm{ext}}(G) = s$ and $r_g(G) = t$.

As for the case $s = 0$, it is similarly proved, by considering $b - 1$ copies of $K_{2,3}$ and a single copy of $K_{2,c}$, where $b = t_1$ and $c = 2t_2 - 4t_1 + 2$. □

Conjecture 2.2 ([54]). Let G be a connected graph such that $r_{\text{ext}}(G) < r_g(G)$. Then, $2r_g(G) < r_{\text{ext}}(G) + 1$.

2.3 Monophonic and *m*-Hull Numbers

For two vertices u and v of a graph G, the *monophonic interval* $J[u,v]$ consists of u,v together with all vertices lying on some chordless $u - v$ path in G. If S is a set of vertices of G, then the *monophonic closure* $J[S]$ consists of S together with all vertices lying on some chordless path joining two vertices of S. If $J[S] = S$, then S is called *m-convex*, and if $J[S] = V(G)$, then S is said to be a *monophonic set*. Similarly to the geodetic case, the smallest m-convex set $[S]_m$ containing S is called the *m-convex hull* of S and an *m-hull set* is a set S such that $[S]_m = V(G)$.

Theorem 2.10 ([83]). *Let $G = (V,E)$ be a graph and $X \subseteq V$ a convex set. Then, X is m-convex if and only if for every pair of nonadjacent vertices $u,v \in X$ and every component of $G - X$ either $V(C) \cap N(u) = \emptyset$ or $V(C) \cap N(v) = \emptyset$.*

Proof. Assume that X is m-convex. The existence of a pair of nonadjacent vertices $u,v \in X$ and a component C of $G - X$ containing a pair u',v' such that $u' \in V(C) \cap N(u)$ and $v' \in V(C) \cap N(v)$ implies the existence of a sequence of vertices $\{w_i\}_{i=0}^{k+1}$ such that $k \geq 2$, $w_0 = u$, $w_1 = u'$, $w_k = v'$, $w_{k+1} = v$, and $\{w_i\}_{i=1}^{k}$ induces a chordless path in C. Hence, there exists a chordless path joining u and v containing at least one vertex outside X, a contradiction.

Conversely, suppose that X in not m-convex. Take a pair of vertices $u,v \in X$ and a chordless $u - v$ path ρ such that $V(\rho) \cap V(G-X) \neq \emptyset$. If $V(\rho) = \{w_i\}_{i=0}^{k+1}$, where $w_0 = u$ and $w_{k+1} = v$, then there exists a pair of indices $r,s \in \{w_i\}_{i=1}^{k}$ such that $r < s$, $w_r \in X$, $w_s \in X$, and $\{w_i\}_{i=r+1}^{s-1} \subseteq V(G-X)$. Hence, w_r, w_s is a pair of nonadjacent vertices in X and there is a component C of $G - X$ such that $V(C) \cap N(w_r) \neq \emptyset$ and $V(C) \cap N(w_s) \neq \emptyset$. □

The *monophonic number* of a graph G, denoted by $m(G)$, is the minimum cardinality of a monophonic set of $V(G)$ [116]. The *m-hull number* of a graph G, denoted by $h_m(G)$, is the minimum cardinality of an m-hull set of $V(G)$.

Certainly, $h_m(G) \leq m(G) \leq g(G)$ and $h_m(G) \leq h(G)$, since every monophonic set is an m-hull set, every geodetic set is monophonic, and every g-hull set is an m-hull set. Nevertheless, it is not true that every g-hull set be monophonic. For example, if $V_1 = \{a,b,c\}$ and $V_2 = \{e,f,g\}$ are the partite sets of the complete bipartite graph $K_{3,3}$, then it is easy to see that the set $W = \{a,b\}$ satisfies $[W]_g = V$ and $J[W] = V \smallsetminus \{c\}$.

At this point, what remains to be done is to ask the following question: *Is there any other general relationship among the parameters $h_m(G)$, $m(G)$, $h(G)$, and $g(G)$,*

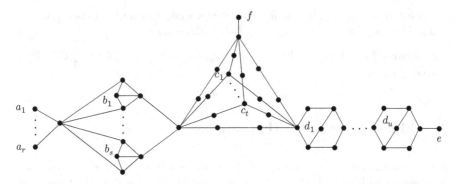

Fig. 2.5 $h_m(G) = r+2, m(G) = r+s+2, h(G) = r+t+2$, and $g(G) = r+t+u+2$

apart from the previous known inequalities? The following realization theorem shows that, unless we restrict ourselves to a specific class of graphs, the answer is negative.

Theorem 2.11 ([119]). *For any integers a, b, c, d such that $3 \leq a \leq b \leq c \leq d$, there exists a connected graph $G = (V,E)$, satisfying one of the following conditions:*

1. $a = h_m(G)$, $b = m(G)$, $c = h(G)$, *and* $d = g(G)$.
2. $a = h_m(G)$, $b = h(G)$, $c = m(G)$, *and* $d = g(G)$.

Proof. Let $G = (V,E)$ be the connected graph shown in Fig. 2.5. We consider the following subsets of vertices: $W_1 = \text{Ext}(G) = \{a_1,\dots,a_r,e,f\}$, $W_2 = W_1 \bigcup \{b_1,\dots,b_s\}$, $W_3 = W_1 \bigcup \{c_1,\dots,c_t\}$, and $W_4 = W_2 \bigcup \{c_1,\dots,c_t\} \bigcup \{d_1,\dots,d_u\}$. Next, we show that W_1 is a minimum m-hull set, W_2 is a minimum monophonic set, W_3 is a minimum g-hull set, and W_4 is a minimum geodetic set.

(i) W_1 is a minimum m-hull set. It is easy to see that every vertex $v \in V \smallsetminus \{b_1,\dots,b_s\}$ lies on some $a_i - e$ monophonic path. Hence, $J[W_1] = V \smallsetminus \{b_1,\dots,b_s\}$. Given a vertex b_i, we take $u_i, v_i \in N(b_i)$ such that $d(v_i, u_i) = 2$. Clearly, $b_i \in J[J[W_1]] = J^2[W_1]$ since $u_i b_i v_i$ is a monophonic path and $u_i, v_i \in J[W_1]$. As a consequence, we have proved that $[W_1]_m = J^2[W_1] = V$. This means that W_1 is an m-hull set of minimum cardinality since every m-hull set must contain all the simplicial vertices of the graph.

(ii) W_2 is a minimum monophonic set. From (i), we immediately conclude that W_2 is a monophonic set. In order to prove that W_2 is a minimum monophonic set, it is enough to remark the following fact. If W is a monophonic set of G, then for $i = 1,\dots,s$ either $b_i \in W$ or $u_i, v_i \in W$ because, in any other case, every path containing b_i must also contain a chord.

(iii) W_3 is a minimum hull set. Notice that:

$$I[W_1] = V \smallsetminus \{b_1,\dots,b_s,c_1,\dots,c_t,d_1,\dots,d_u\} \cup N(c_1) \cup \dots \cup N(c_t),$$

$$I[I[W_1]] = I^2[W_1] = V \smallsetminus \{c_1,\dots,c_t\} \cup N(c_1) \cup \dots \cup N(c_t) = [\text{Ext}(G)]_g.$$

Next, observe that every g-hull set W satisfies, first, $Ext(G) \subseteq W$ and, second, $W \cap (c_i \cup N(c_i)) \neq \emptyset$, for every $i = 1, \ldots, t$. Hence, W_3 is a g-hull set of minimum cardinality.

(iv) W_4 is a minimum geodetic set. It is easy to see that every vertex $v \in V$, v lies on some $a_i - e$ geodesic joining two vertices of W_4. By the other hand, if we consider W as a geodetic set, then (a) either $b_i \in W$ or $u_i, v_i \in W$, $i = 1, \ldots, s$; (b) either $c_j \in W$ or $z_j \in W$ for some $z_j \in N(c_j)$, $j = 1, \ldots, t$; and (c) either $d_h \in W$ or $N(d_h) \subset W$, $h = 1, \ldots, u$. Hence, W_4 is a minimum geodetic set.

As a consequence, we have proved that $h_m(G) = r+2$, $m(G) = r+s+2$, $h(G) = r+t+2$, and $g(G) = r+s+t+u+2$, from which both statements of the theorem immediately follow. □

2.4 Convexity Number

The *convexity number*[1] of a connected graph G, denoted by $con(G)$, is the maximum cardinality of a proper convex set of G [48,64]. The *clique number* $\omega(G)$ of a graph G is the maximum order of a clique in G. The *independence number* $\alpha(G)$ is the cardinality of a largest independent set, i.e., $\alpha(G) = \omega(\overline{G})$. Every non-complete graph G satisfies $2 \leq \omega(G) \leq con(G) \leq n-1$, as the vertex set of every clique is convex. Moreover, $con(G) = n-1$ if and only if G contains at least a simplicial vertex v since, as was shown in Proposition 2.2, $V(G) - v$ is a convex set if and only if $N(v)$ induces a clique.

Theorem 2.12 ([36]). *Let G be a graph of diameter* $diam(G) = 2$. *If its complement $\overline{G}$ is not connected and at least two of the components of $\overline{G}$ are nontrivial, then* $con(G) = \omega(G)$.

Proof. Let $\{H_i\}_{i=1}^k$ de components of $\overline{G}$. Certainly, $G = \vee_{i=1}^k \overline{H}_i$. Let C be a maximum proper convex set in G. Let $S_i = C \cap V(H_i)$, for every $i \in [k]$. To end the proof, it is enough to show that if $S_i \neq \emptyset$, then S_i induces a clique in $\overline{H}_i$. Suppose, to the contrary, that $S_1 \neq \emptyset$ and there exist two distinct vertices $u, v \in S_1$ such that $d_G(u,v) = 2$. Hence, $V(G) \setminus V(\overline{H}_1) \subseteq I_G[u,v] \subseteq C$. By hypothesis, there exists an index $j \neq 1$ such that H_j is nontrivial. Since H_i is connected, there exist two distinct vertices $a, b \in V(H_j)$ which are adjacent in H_j, i.e., a and b are not adjacent in $\overline{H}_j$. Hence, $d_G(a,b) = 2$ and both a and b are vertices in C adjacent to all vertices of $\overline{H}_1$, which means that $C = V(G)$, a contradiction. □

The converse of the preceding theorem is not always true. Take, for example, de graph $G \cong K_2 \vee \overline{K}_2$, and check that $diam(G) = 2$, $\overline{G} = \overline{K}_2 + K_2$, and $con(G) = \omega(G) = 3$.

---

[1] With respect to the geodesic convexity.

Fig. 2.6 $(K_2 + \overline{K}_3) \vee \overline{K}_2$

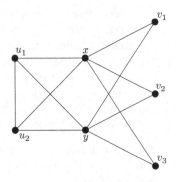

It remains an *open problem* to characterize the family of graphs of diameter 2 for which the convexity number and the clique number are equal.

A graph G of order n such that $\omega(G) = l$ and $\mathrm{con}(G) = k$ is called an (l,k,n)-graph. The *clique ratio* and *convexity ratio* of an (l,k,n)-graph G are defined respectively as $r_\omega(G) = \frac{\omega(G)}{n}$ and $r_{\mathrm{con}}(G) = \frac{\mathrm{con}(G)}{n}$.

Theorem 2.13 ([64]). *For every triple l,k,n of integers with $2 \le l \le k \le n-1$, there exists a non-complete (l,k,n)-graph.*

Proof. We consider two cases.

Case 1: $l = k$. For $k = 2$ and $k = n-1$, the graphs $K_{2,n-2}$ and $K_n - e$, where $e \in E(K_n)$, have, respectively, the desired properties. So, we assume that $3 \le k \le n-2$. Consider the graph $G = (K_{k-1} + \overline{K}_{n-k-1}) \vee \overline{K}_2$ (in Fig. 2.6 the case $n = 7$ and $k = 3$ is shown). Clearly, G is a connected graph of order n and $\omega(G) = k \le \mathrm{con}(G)$. Let S be a maximum proper convex set of G. If $V(\overline{K}_2) = \{x,y\}$, then $|S \cap \{x,y\}| \le 1$, since $I[x,y] = V(G)$. We claim that $S \cap V(\overline{K}_{n-k-1}) = \emptyset$. Assume, to the contrary, that this is not the case. First assume that S contains at least two vertices, say v_1, v_2, of $V(\overline{K}_{n-k-1})$. Then, $x,y \in I[v_1,v_2]$, and so $I[S] = V(G)$, a contradiction. Hence, S contains exactly one vertex of $V(\overline{K}_{n-k-1})$, say v_1. Since $k \ge 3$, it follows that S contains at least two distinct vertices $u,v \notin V(\overline{K}_{n-k-1})$. We may assume w.l.o.g. that $u \notin \{x,y\}$, as $|S \cap \{x,y\}| \le 1$. Since x and y lie on a $u - v_1$ geodesic, it follows that $x,y \in I[u,v_1]$, again a contradiction. Hence $S \cap V(\overline{K}_{n-k-1}) = \emptyset$, as claimed. Because S contains at most one of x and y, $\mathrm{con}(G) = |S| \le k$ and so $\mathrm{con}(G) = k$.

Case 2: $l < k$. Take the graph $F = (K_{l-1} + \overline{K}_{k-l-1}) \vee \overline{K}_2$, where $V(K_{l-1}) = \{u_i\}_{i=1}^{l-1}$, $V(\overline{K}_{k-l-1}) = \{v_i\}_{i=1}^{k-l-1}$, and $\overline{K}_2 = \{x,y\}$. We consider two cases.

Subcase 2.1: $n - 3 \le k \le n - 1$. If $k = n - 1$, let F_1 be the graph obtained from F by adding a new vertex r and the pendant edge xr. If $k = n - 2$, let F_2 be the graph obtained from F by adding two new vertices r,s and the edges xr, rs, sy. If $k = n - 3$, let F_3 be the graph obtained from F by adding three new vertices r,s,t and the edges xr, rs, st, ty. It is a routine exercise to check that for $1 \le i \le 3$, F_i is a graph of order n with $\omega(F_i) = \omega(F) = l$ and $\mathrm{con}(F_i) = k$.

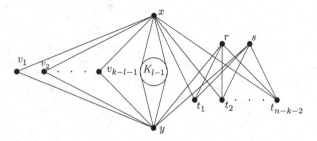

Fig. 2.7 Graph G of order $n \geq k+4$ with $\omega(G) = l$ and $\mathrm{con}(G) = k$

Subcase 2.2: $k \leq n-4$. Take the graphs F and $H = K_{2,n-k-2}$. Let G be the graph obtained from F and H by adding the edges $\{xt_i\}_{i=1}^{n-k-2} \cup \{yr, ys\}$ (see Fig. 2.7). Clearly, G is a graph of order n with $\omega(G) = l$. Observe that $V(F)$ is a convex set of G of order k. It remains thus only to show that $V(F)$ is a maximum proper convex of G. Assume, to the contrary, that S' is proper convex set of G with $|S'| \geq k+1$. Hence, $S' \cap V(H) \neq \emptyset$. Notice that $[r,s] = V(G)$ and, for every pair of distinct indexes $i,j \in \{1,\ldots,n-k-2\}$, $[t_i,t_j] = V(G)$. Thus, $1 \leq |S' \cap V(H)| \leq 2$, i.e., $k+1 \leq |S'| \leq k+2$ and $k-1 \leq |S' \cap V(F)| \leq k$. If we suppose that $x \in S'$ (resp. t_1), then $r \notin S'$ (resp. $s \notin S'$), as $[x,r] = V(G)$ (resp. $[y,t_1] = V(G)$). In other words, in all cases we arrive at a contradiction. □

Corollary 2.2 ([61]). *For every pair s,t of rational numbers with $0 < s \leq t < 1$, there exists a graph with $r_\omega(G) = s$ and $r_{\mathrm{con}}(G) = t$.*

In the proof of Theorem 2.13, all the (l,k,n)-graphs constructed for $l \geq 3$ have the properties that there are exactly two maximum cliques, a unique maximum convex set S, and all vertices of both maximum cliques are in S. At this point, two natural questions arise: (1) Do (l,k,n)-graphs with arbitrarily many maximum convex sets exist? (2) Do (l,k,n)-graphs in which no maximum clique set is contained in any maximum convex set exist? In [61], both questions were partially answered.

Theorem 2.14 ([61]). *Let s,t be rational numbers with $0 < s < t < 1$ and let M be a positive integer. Then,*

1. *If $t \leq \frac{1}{2}$, then there exists a graph G with $r_\omega(G) = s$ and $r_{\mathrm{con}}(G) = t$, such that G contains at least M distinct maximum convex sets.*
2. *If $s+t < 1$ and $t > \frac{1}{2}$, then there exists a graph G with $r_\omega(G) = s$ and $r_{\mathrm{con}}(G) = t$, such that if Ω is the set of vertices of a maximum clique in G and S is a maximum convex set in G, then $d(\Omega,S) = \min\{d(u,v) : u \in \Omega, v \in S\} \geq M$.*

Sketch of proof.

1. Take $s = \frac{a}{b}$ and $t = \frac{c}{d}$ such that $0 < s+t < 1$ and $t \leq \frac{1}{2}$, where a,b,c,d are integers such that $0 < a < b$ and $0 < c < d$. Take $p = r(bc - ad)$ and $q = r(bd +$

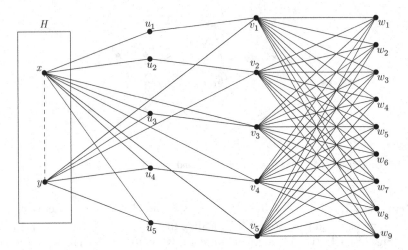

Fig. 2.8 A graph G with $r_\omega(G) = 1/3$ and $r_{\mathrm{con}}(G) = 1/2$. Note that $V(H) \cup \{u_3, u_4, u_5, v_3\}$, $V(H) \cup \{u_3, u_4, u_5, v_4\}$, and $V(H) \cup \{u_3, u_4, u_5, v_5\}$ are maximum convex sets

$ad - 2bc) - 1$, where $r = \lceil \max\{5, M+2\} \rceil$. Check that $p \geq \max\{5, M+2\}$ and $q \geq 4$. Consider the graph $H \cong K_{\mathrm{rad}+1} - e$ where $e = xy \in E(K_{\mathrm{rad}+1})$, the complete bipartite graph $F \cong K_{p,q}$ and the null graph $K \cong \overline{K}_p$.

Construct the graph $G = (V, E)$ such that $V(G) = V(H) \cup V(F) \cup V(K)$ and $E = E(H) \cup E(F) \cup E(K) \cup \{u_i v_i\}_{i=1}^p \cup \{xv_i, yu_i\}_{i=3}^p \cup \{yv_1, yv_2\}$, where $V(K) = \{u_i\}_{i=1}^p$ and $\{v_i\}_{i=1}^p$ is the stable set of F of order p. In Fig. 2.8, the graph G when $s = 1/3, r = 1/2$, and $M = 3$ is displayed.

Observe that G is a graph of order rbd and clique number $\omega(G) = \mathrm{rad}$. Check that, for $i = 3, \ldots, p$, the set $S_i = V(H) \cup \{u_3, \ldots, u_p, v_i\}$ is a convex set of G. Thus, $\mathrm{con}(G) \geq |S_i| = rbc$. Next, notice that every convex set of G cannot contain a pair of nonadjacent vertices of H. Finally, check that every set of cardinality at least $rbc + 1$ is not convex.

2. Take $s = \frac{a}{b}$ and $t = \frac{c}{d}$ such that $0 < s + t < 1$ and $t > \frac{1}{2}$, where a, b, c, d are integers such that $0 < a < b$ and $0 < c < d$. Choose an integer $r \geq 5M$ sufficiently large so that $t > \frac{1}{2} + \frac{1}{rbd}$. Consider the graph $H_0 \cong K_{\mathrm{rad}+1} - e$ where $e = x_0 y_0 \in E(K_{\mathrm{rad}+1})$. Let $h_1, h_2, \ldots, h_l$ be integers such that $3 \leq h_i \leq 5$ for every $i \in \{1, \ldots, l\}$ and $\sum_{i=1}^l h_i = r(bd - bc - ad) \geq r$. Since $r \geq 5M$, it follows that $l \geq r/5 \geq M$. For every $i \in \{1, \ldots, l\}$, consider the graph $H_i \cong K_{h_i} - e_i$, where $e_i = x_i y_i \in E(K_{h_i})$. Take the graph $F \cong K_{2,q}$, where $q = rbc - 3$.

Construct the graph $G = (V, E)$ such that $V(G) = \cup_{i=0}^l V(H_i) \cup V(F)$ and $E = \cup_{i=0}^l E(H_i) \cup E(F) \cup \{x_i x_{i+1}, y_i x_{i+1}, x_i y_{i+1}, y_i y_{i+1}\}_{i=0}^{l-1} \cup \{x_l u_1, y_l u_1, x_l u_2, y_l u_2\}$, where $\{u_1, u_2\}$ is the stable set of F of order 2 (see Fig. 2.9).

Observe that G is a graph of order rbd and clique number $\omega(G) = \mathrm{rad}$. Notice that G has two maximum cliques, namely, $\Omega_1 = V(H_0) \setminus \{x_0\}$ and $\Omega_2 = V(H_0) \setminus \{y_0\}$. Check that $S = V(F) \cup \{x_l\}$ is the unique maximum convex set of G. Finally, observe that $d(\Omega_1, S) = d(\Omega_2, S) = l \geq M$.   □

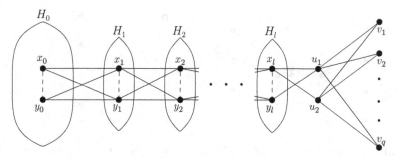

Fig. 2.9 A graph G such that $r_\omega(G) = s$ and $r_{con}(G) = t$, where $s + t < 1$ and $t > \frac{1}{2}$

Table 2.2 Clique and convexity numbers of some basic graph families

G^a	P_n	C_{2l}	C_{2l+1}	T_n	K_n	$K_{p,q}^b$	W_n^c	Q_n
$\omega(G)$	2	2	2	2	n	2	3	2
$con(G)$	$n-1$	l	$l+1$	$n-1$	$n-1$	2	$n-2$	2^{n-1}

$^aG \not\cong K_1$
$^b2 \le p \le q$
$^cn \ge 5$

Question 2.1 ([61]). Are conditions $t \le \frac{1}{2}$ and $t > \frac{1}{2}$ of Theorem 2.14 needed?

A graph G is called *polyconvex* if for every integer i with $1 \le i \le con(G)$, there exists a convex set of cardinality i in G. All the graph families displayed in Table 2.2 are formed by polyconvex graphs, except the family of hypercubes [48].

Proposition 2.3 ([48]). *For $n \le 2$, a set S in the n-dimensional hypercube Q_n is convex if and only if S induces a k-dimensional hypercube Q_k in Q_n, for some integer k with $0 \le k \le n$. Hence, for every $n > 2$, Q_n is not polyconvex.*

Proof. If S is a set of vertices in Q_n that induces a Q_i for some $i \in [n]$, then S is convex in Q_n. So it remains only to prove the converse. We proceed by induction on n. The result is certainly true for $n = 2$. Assume that every convex set in Q_k ($k \ge 2$) induces a Q_i for some $i \in [k]$.

Let Q_{k+1} be a $(k+1)$-cube formed from two copies G_1 and G_2 of Q_k, i.e., $Q_{k+1} = Q_k \Box K_2$, $V(G_1) = Q_k \times \{1\}$ and $V(G_2) = Q_k \times \{2\}$. Let S be a convex set in Q_{k+1}. If either $S = V(Q_{k+1})$ or $S \subseteq V(G_i)$ for $i = 1$ or $i = 2$, then the result holds. Thus we may assume that $S = S_1 \cup S_2$, where $S_1 = S \cap V(G_1)$ and $S_2 = S \cap V(G_2)$ are both nonempty. Certainly S_i is convex in G_i, for $i = 1, 2$. Hence, by induction hypothesis, $G[S_1] = Q_r$ and $G[S_2] = Q_s$, where $0 \le r, s \le k$. Take $u_1 \in S_1$ and $v_2 \in S_2$. If u_2 is the neighbor of u_1 in G_2 and v_1 is the neighbor of v_2 in G_1, then $\{u_2, v_1\} \subseteq I[u_1, v_2]$. Hence, $r = s$ and $S = Q_{r+1}$.   □

Theorem 2.15 ([48]). *Given any pair of integers n, k with $2 \le k \le n-1$, there exists a polyconvex connected graph of order n with $con(G) = k$.*

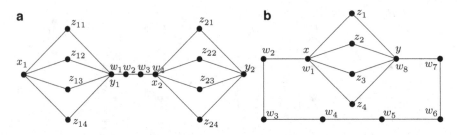

Fig. 2.10 (a) Graph G s.t. $n = 14$ and $con(G) = k = 10$, (b) Graph G s.t. $n = 12$ and $con(G) = k = 8$

Proof. For $k = n-1$, take $G = K_n$. So we may assume that $2 \le k \le n-2$. Consider the graph $G = (K_{k-1} + \overline{K}_{n-k-1}) \vee \overline{K}_2$ (in Fig. 2.6 the case $n = 7$ and $k = 3$ is shown). In the proof of Theorem 2.13 it was shown that G is a graph of order n with $con(G) = k$. Finally, if $V(K_{k-1}) = \{u_i\}_{i=1}^{k-1}$ and $V(\overline{K}_2) = \{x, y\}$, notice first that $\{u_i\}_{i=1}^{k-1} \cup \{x\}$ induces a clique and hence it is a convex set of order k, and, second, that, for every $j \in \{1, \dots, k-1\}$, $\{u_i\}_{i=1}^{j}$ induces a clique and thus it is a convex set of order j. $\square$

Theorem 2.16 ([124]). *Given any pair of integers n, k with $2 \le k \le n-1$, there exists a polyconvex connected K_3-free graph of order n with $con(G) = k$.*

Proof. For $k = 2$, take $G = K_{2,n-2}$. So we may assume that $3 \le k \le n-1$. We consider three cases.

Case 1: $k+1 \le n \le \frac{3}{2}k - \frac{1}{2}$. If $3 \le k \le 4$, P_4 and P_5. Suppose thus that $k \ge 5$. Take two copies G_1 and G_2 of $K_{2,n-k}$ and a copy G_3 of P_{2k-n-2}. Let G be the graph obtained from G_1, G_2, and G_3 by identifying the vertices w_1 and w_{2k-n-2} with y_1 and x_2, respectively, where $V(G_i) = \{x_i, y_i\} \cup \{z_{ij}\}_{j=1}^{n-k}$ and $V(G_3) = \{w_i\}_{i=1}^{2k-n-2}$ (in Fig. 2.10a the case $n = 14$ and $k = 10$ is shown). Notice (1) G is a K_3-graph of order n, (2) $2k - n - 2 \ge 0$, and (3) if $n = k+1$, then G is P_n. We form convex sets of orders 1 through $2k - n - 2$ by taking a subpath of appropriate length from G_3. We form convex sets of orders $2k - n - 1$ and $2k - n$ by taking all the vertices of G_3 and adding at most one vertex from each of the partite classes of order $n - k$ of G_1 and G_2. We form a convex set of orders $n - k + 2$ to $k - 1$ by taking all the vertices of G_1 together with an appropriate number of adjacent vertices of G_3. Finally, we form a convex set of order k by taking all the vertices of G_1 and G_3 and one more vertex from the partite class of order $n - k$ of G_2. Since $(n - k + 2) - (2k - n) \le 1$, we have found convex sets of all orders between 1 and $k = con(G)$.

Case 2: $\frac{3}{2}k \le n \le 2k - 1$. Take a copy G_1 of $K_{2,2k-n}$ and a copy G_2 of P_{2n-2k}. Let G be the graph obtained from G_1 and G_2 by identifying the vertices w_1 and w_{2n-2k} with x and y, respectively, where $V(G_1) = \{x, y\} \cup \{z_i\}_{i=1}^{2k-n}$ and $V(G_2) = \{w_i\}_{i=1}^{2n-2k}$ (in Fig. 2.10b the case $n = 12$ and $k = 8$ is shown). Notice (1) G is a K_3-graph of order n, (2) $2k - n \ge 1$ and (3) if $n = 2k - 1$ then G is C_n. We form convex sets of orders 1 through $n - k + 1$ by taking a subpath of appropriate length from G_2. We form convex sets of orders $2 + 2k - n$ through k by taking all the vertices of G_1 and adding suitable numbers of adjacent vertices of G_2. Since $(2 + 2k - n) - (n - k + 1) \leq 1$, we have found convex sets of all orders between 1 and $k = \text{con}(G)$.

Case 3: $2k \leq n$. Let $m = \lfloor \frac{n-2k+4}{2} \rfloor$. Note that $m \geq 2$ since $n \geq 2k$. Start with a copy G_1 of $K_{m,m}$ or $K_{m,m+1}$, depending on whether $n - 2k + 4$ is even or odd. Let G be the graph obtained by replacing one edge of G_1 with a set Ω of $k - 2$ paths of length 3. Let v be one of the vertices of Ω of degree greater than 2. Let S be the set of vertices consisting of v, all the neighbors of v in Ω and one more neighbor of v. Then S is a (maximum) convex set of order k. Convex sets of all smaller orders occur as subsets of S containing v. Notice that when $k = 3$ and $n = 7$, then this construction does not work. For this case, take the graph $K_{3,3}$ and replace one edge with a path of length 2. Then the three vertices of this path form a largest convex set.   □

Finally, the following Nordhaus–Gaddum type result has been obtained.

Theorem 2.17 ([48, 104]). $2 \log_4 n \leq \text{con}(G) + \text{con}(\overline{G}) \leq 2n - 2$

Proof. The upper bound is obvious since for every graph G, $\text{con}(G) \leq n - 1$. To prove the lower bound we use the well-known result that, for every pair of positive integers h, k, the Ramsey number $R(h, k)$ is bounded above by $\binom{h+k-2}{h-1}$ (see Corollary 12.18 of [71]). In particular, $R(h, k) \leq 2^{h+k-2}$. Notice that if G is a graph of order n, then $R(\omega(G) + 1, \omega(\overline{G}) + 1) > n$. Suppose that there is some graph G of order n such that $2 \log_4 n > \text{con}(G) + \text{con}(\overline{G})$. Hence, $R(\omega(G) + 1, \omega(\overline{G}) + 1) \leq 2^{\omega(G)+\omega(\overline{G})} \leq 2^{\text{con}(G)+\text{con}(\overline{G})} < 2^{2\log_4 n} = n$, a contradiction.   □

Further results involving ideas, concepts, and invariants appearing in Sects. 2.1–2.4 can be found in [16, 19, 22, 48, 51, 54, 57, 59, 61, 63–65, 69, 72, 73, 75, 82, 98, 99, 101, 113, 119, 124, 154, 157, 160, 161, 163, 170].

2.5 Forcing Geodomination

Let S be a minimum geodetic set of a graph G. A subset T of S is called a *g-forcing subset* of S, if S is the unique minimum geodetic set that contains T. The minimum size of a forcing subset of a minimum geodetic set in G is called the *forcing geodetic number* $f_g(G)$ of G.

Let S be a minimum hull set of a graph G. A subset T of S is called an *h-forcing subset* of S, if S is the unique minimum hull set that contains T. The minimum size of an *h*-forcing subset of a minimum hull set in G is called the *forcing hull number* $f_h(G)$ of G.

As an immediate consequence of these definitions, the following properties hold.

Proposition 2.4. *Let G be a connected graph G. Then,*

- $f_g(G) \leq g(G)$ *and* $f_h(G) \leq h(G)$.

Table 2.3 Forcing and standard parameters of some basic graph families

G^a	P_n	C_{2l}	C_{2l+1}	T_n^k	K_n	$K_{p,q}{}^b$	Q_n
$f_h(G)$	0	1	0	0	0	2	1
$h(G)$	2	2	3	k	n	2	2
$f_g(G)$	0	1	0	0	0	4	1
$g(G)$	2	2	3	k	n	4	2
$f_c(G)$	1	2	2	$k-1$	$n-1$	2	2
$con(G)$	$n-1$	l	$l+1$	$n-1$	$n-1$	2	2^{n-1}

$^a G \not\cong K_1$
$^b 5 \leq p \leq q$

- $f_g(G) = 0$ (resp. $f_h(G) = 0$) if and only if G has a unique minimum geodetic set (resp. hull set).
- $f_g(G) = 1$ (resp. $f_h(G) = 1$) if and only if G has at least two minimum geodetic sets (resp. hull sets) and there exists a vertex belonging to exactly one minimum geodetic set (resp. hull set).
- $f_g(G) \geq 2$ (resp. $f_h(G) \geq 2$) if and only if G each vertex of each minimum geodetic set (resp. hull set) belongs to a least one more minimum geodetic set (resp. hull set).
- There are graphs G satisfying $f_g(G) < f_h(G)$. For example, if $2 < s$, then $f_h(K_{2,s}) = 1$ and $f_g(K_{2,s}) = 0$.

In Table 2.3, both the forcing geodetic number (resp. forcing hull number) and the geodetic number (resp. hull number) of some basic graphs are displayed.

Theorem 2.18 ([49]). *Given any pair of integers a,b with $0 \leq a \leq b$, $2 \leq b$ and $(a,b) \neq (2,2)$, there exists a graph G such that $f_g(G) = a$ and $g(G) = b$.*

Proof. First check that if $g(G) = 2$, then $f_g(G) < 2$. If $a = 0$, take $G \cong K_b$. Next, suppose that $0 < a < b$. We distinguish three cases.

Case 1: $a = 1$. If $b = 2$, take any even cycle. Suppose that $b \geq 3$. Consider the graph G_1 in Fig. 2.11. Notice that $\{u_1, \ldots, u_{b-2}, v_3, x\}$ is a minimum geodetic set of G_1 and thus $g(G_1) = b$. To prove that $f_g(G_1) = 1$, notice first that $X = \{u_1, \ldots, u_{b-2}, v_3, x\}$ and $Y = \{u_1, \ldots, u_{b-2}, v_3, y\}$ are two distinct minimum geodetic sets and second that X is the unique minimum geodetic set containing $\{x\}$.

Case 2: $a > 1$ and $b = a+1$. Consider the graph G_2 in Fig. 2.11. Notice that $\{u_2, \ldots, u_{a+1}, v_1\}$ is a minimum geodetic set of G_2 and thus $g(G_2) = a+1$. Let W be a minimum geodetic set of G_2. Since $W \cap \{u_i\}_{i=1}^{a+1} \neq \emptyset$, $W \cap \{v_i\}_{i=1}^{a+1} \neq \emptyset$ and, for every $i \in \{1, \ldots, a+1\}$, $W \cap \{u_i, v_i\} \neq \emptyset$, it follows that W is not the unique minimum geodetic set containing any of its subsets W' with $|W'| < a$. Thus $f_g(G_2) \geq a$. On the other hand, $W = \{u_2, \ldots, u_{a+1}, v_1\}$ is the unique minimum geodetic set containing $\{u_2, \ldots, u_{a+1}\}$, which means that $f_g(G_2) = a$.

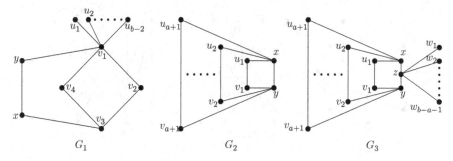

Fig. 2.11 Case 1: $f_g(G_1) = 1$, $g(G_2) = b$. Case 2: $f_g(G_2) = g(G_2) - 1 = a$. Case 3: $f_g(G_3) = a$, $g(G_3) = b$

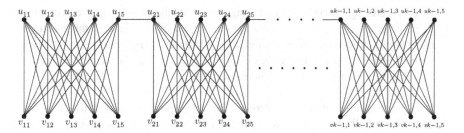

Fig. 2.12 Graph G_{2k} with $f_g(G_{2k} = g(G_{2k}) = a$

Case 3: $a > 1$ and $b \geq a + 2$. Consider the graph G_3 in Fig. 2.11. Notice that $\{u_2, \ldots, u_{a+1}, v_1, w_1, \ldots, w_{b-a-1}\}$ is a minimum geodetic set of G_3 and thus $g(G_3) = b$. Next, notice that every minimum geodetic set of G_3 has the form $S = U \cup V \cup W$ where $U \subseteq \{u_i\}_{i=1}^{a+1}$, $V \subseteq \{v_i\}_{i=1}^{a+1}$, and $W = \{w_i\}_{i=1}^{b-a-1}$, with $U \neq \emptyset$ and $V \neq \emptyset$. This implies that if S is a minimum geodetic set of G_3, then it is not the unique minimum geodetic set containing any of its subsets W' with $|W'| < a$. Thus $f_g(G_3) \geq a$. On the other hand, $\{u_2, \ldots, u_{a+1}, v_1, w_1, \ldots, w_{b-a-1}\}$ is the unique minimum geodetic set containing $\{u_2, \ldots, u_{a+1}\}$, which means that $f_g(G_3) = a$.

Finally, assume that $3 \leq a = b$. We distinguish three cases.

Case 1: $a = 2k$ where $k \geq 2$. For $k = 2$, take $K_{5,5}$. Suppose that $k \geq 3$. Consider the graph G_{2k} obtained from $k - 1$ copies $F_1, \ldots, F_{k-1}$ of $K_{5,5}$ by adding $k - 2$ edges, namely, the set $\{u_{i5}u_{i+1,1}\}_{i=1}^{k-2}$, where, for every $i \in \{1, \ldots, k-1\}$, $V(F_i) = U_i \cup V_i = \{u_{ij}\}_{j=1}^{5} \cup \{v_{ij}\}_{j=1}^{5}$ (see Fig. 2.12). Notice that every minimum geodetic set of G_{2k} contains exactly one vertex from each of $U_1 \setminus \{u_{15}\}$ and $U_{k-1} \setminus \{u_{k-1,1}\}$ and exactly two vertices from each set V_i, $1 \leq i \leq k - 1$. Therefore, $f_g(G_{2k}) = g(G_{2k}) = 2k = a$.

Case 2: $a = 2k + 1$ where $k \geq 1$. For $k = 1$, take the graph H in Fig. 2.13. For $k = 2$, take the graph $C_5 \circ \overline{K}_5$. Suppose that $k \geq 3$. Let F_i' the graph obtained from

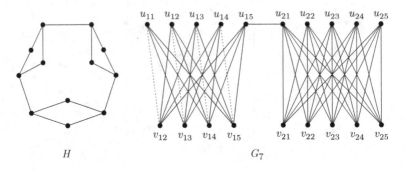

Fig. 2.13 $f_g(H) = g(H) = 3$ and $f_g(G_7) = g(G_7) = 7$

$K_{5,4}$ by deleting a maximum matching. Consider the graph G_{2k+1} constructed from the graph G_{2k+1} in Case 1 by replacing F_1 by F_1' (in Fig. 2.13 the graph G_7 is depicted). Notice that every minimum geodetic set of G_{2k} contains exactly two vertices from $U_1 \setminus \{u_{15}\}$, one vertex from $U_{k-1} \setminus \{u_{k-1,1}\}$ and exactly two vertices from each set V_i, $1 \le i \le k-1$. Therefore, $f_g(G_{2k}) = g(G_{2k}) = 2k+1 = a$. □

In a quite similar way it is proved the corresponding realization theorem involving both hull parameters. Notice that, in this case, the equality $(a,b) = (2,2)$ is possible, as, for example, $f_h(K_{3,3}) = h(K_{3,3}) = 2$.

Theorem 2.19 ([53]). *Given any pair of integers a,b with $0 \le a \le b$ and $2 \le b$, there exists a graph G such that $f_h(G) = a$ and $h(G) = b$.*

Lemma 2.1 ([171]). *Let m,t be positive integers with $m \ge 2$. Let $\{G_i\}_{i=1}^m$ be a family of m disjoint connected graphs of order at least $t+1$, such that each of them contains a t-set S_i such that $N_{G_i}[S_i]$ induces a clique. Let G be the graph obtained from $\{G_i\}_{i=1}^m$ by identifying $S_1, S_2, \ldots, S_m$. Then,*

(a) $h(G) = \sum_{i=1}^m h(G_i) - mt.$

(b) $g(G) = \sum_{i=1}^m g(G_i) - mt.$

(c) $f_h(G) = \sum_{i=1}^m f_h(G_i).$

(d) $f_g(G) = \sum_{i=1}^m f_g(G_i).$

Sketch of proof. To prove (a) and (b), check that a set T is a minimum hull (resp. geodetic) set of G if and only if $(T \cap V(G_i)) \cup S_i$ is a minimum hull (resp. geodetic) set of G_i for $i = 1, 2, \ldots, m$.

To prove (c) and (d) check that a set F is a forcing hull (resp. geodetic) set of G if and only if $F \cap V(G_i)$ is a forcing hull (resp. geodetic) set of G_i for $i = 1, 2, \ldots, m$.

□

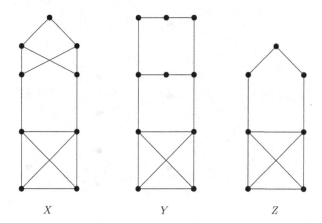

Fig. 2.14 $h(X) = h(Y) = h(Z) = 3, g(X) = g(Z) = 3, g(Y) = 4$

Theorem 2.20 ([171]). *Given any pair of nonnegative integers a,b with $a+b \geq 2$, there exists a two-connected graph G such that $f_h(G) = a$, $f_g(G) = b$, $h(G) = a + b + c$, and $g(G) = a + 2b + c$.*

Sketch of proof. Take a collection $\{X_i\}_{i=1}^{a}$ of disjoint graphs isomorphic to the graph X of Fig. 2.14. Take a collection $\{Y_i\}_{i=1}^{b}$ of disjoint graphs isomorphic to the graph Y of Fig. 2.14. Take a collection $\{X_Z\}_{i=1}^{c}$ of disjoint graphs isomorphic to the graph Z of Fig. 2.14. Observe that $\mathrm{Ext}(X) \cong \mathrm{Ext}(Y) \cong \mathrm{Ext}(Z) \cong K_2$. Let G be the graph obtained from these three families by identifying all the extreme sets. Hence, according to Lemma 2.1, $h(G) = 3a + 3b + 3c - 2(a+b+c) = a+b+c$, $h(G) = 3a + 4b + 3c - 2(a+b+c) = a+2b+c$, $f_h(G) = a$, and $f_g(G) = b$, since

- $h(X) = 3, g(X) = 3, f_h(X) = 1, f_g(X) = 0$.
- $h(Y) = 3, g(Y) = 4, f_h(Y) = 0, f_g(Y) = 1$.
- $h(Z) = 3, g(Z) = 3, f_h(Z) = 0, f_g(Z) = 0$. $\qquad\square$

Theorem 2.21 ([171]). *Given any pair of nonnegative integers a,b, there exists a two-connected graph G such that $f_h(G) = a$ and $f_g(G) = b$.*

Proof. If $a = b = 0$, take any extreme geodesic graph. If $a = 1$ and $b = 0$, take the graph X in Fig. 2.14. If $a = 0$ and $b = 1$, take the graph Y in Fig. 2.14. If $a + b \geq 1$, make use of Theorem 2.20. $\qquad\square$

Conjecture 2.3 ([171]). For any integers a,b,c,d with $a \leq c \leq d$, $b \leq d$, and $c \geq 2$, there exists a connected graph G with $f_h(G) = a$, $f_g(G) = b$, $h(G) = c$, and $g(G) = d$.

Let S be a maximum proper convex set of a graph G. A subset T of S is called a *forcing* subset of S, if S is the unique maximum proper convex set that contains T. The minimum size of a forcing subset of a maximum proper convex set in G is called the *forcing convexity number* $f_c(G)$ of G.

As an immediate consequence of these definitions, the following properties hold.

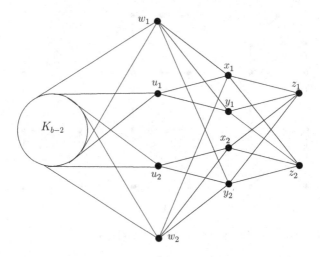

Fig. 2.15 $h(X) = h(Y) = h(Z) = 3$, $g(X) = g(Z) = 3$, $g(Y) = 4$

Proposition 2.5. *Let G be a connected graph G. Then,*

- $f_c(G) \leq g(G)$.
- $f_c(G) = 0$ *if and only if G has a unique maximum proper convex set.*
- $f_c(G) = 1$ *if and only if G has at least two maximum proper convex sets and there exists a vertex belonging to exactly one maximum proper convex set.*
- $f_c(G) \geq 2$ *if and only if G each vertex of each maximum proper convex set belongs to a least one more minimum geodetic set.*
- *There is no graph G satisfying* $(f_g(G), \mathrm{con}(G)) = (0, 2)$.

Proposition 2.6. *Let G be a connected graph G with* $\mathrm{con}(G) = 2$. *If* $G \not\cong P_3$, *then* $f_c(G) = 2$.

Proof. Since $\mathrm{con}(G) = 2$, every pair of adjacent vertices forms a maximum convex set. If G contains a leaf, then either $G \cong P_3$ in which case $f_c(G) = 1$ or G has order at least 4, in which case $\mathrm{con}(G) = n - 1 \geq 3$. Hence, every vertex of G belongs to at least two maximum convex sets. □

Theorem 2.22 ([52]). *Given any pair of integers a, b with* $0 \leq a \leq b$ *and* $3 \leq b$, *there exists a graph G such that* $f_c(G) = a$ *and* $\mathrm{con}(G) = b$.

Sketch of proof. If $3 \leq a = b$, take K_{b+1}. If $1 \leq a < b$, take a tree T_{b+1}^{a+1}, with $b + 1$ vertices and $a + 1$ leaves. Suppose hence that $a = 0$ and $3 \leq b$. Let G be the graph depicted in Fig. 2.15. Observe that G is a graph of order $b + 8$, diameter 3, and radius 2. Check that $V(K_{b-2}) \cup \{u_1, u_2\}$ is the unique maximum proper set of G. □

Conjecture 2.4. For any integers a, n, d with $3 \leq a \leq n - 1$ and $2 \leq d$, there exists a graph G of order n and diameter d such that $f_c(G) = 0$ and $\mathrm{con}(G) = a$.

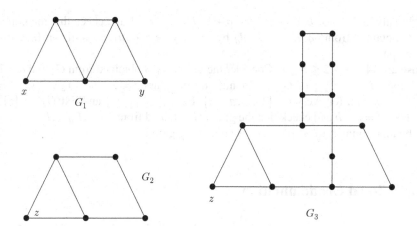

Fig. 2.16 G_1, G_2, and G_3 are $(1,1,3)$-, $(1,2,3)$-, and $(1,3,4)$-graphs, respectively

Let S be a minimum geodetic set of a graph G. A forcing subset T is called *critical* if no proper subset of T is a forcing subset of S. The maximum order of a critical forcing subset of a minimum geodetic set in G is called the *upper forcing geodetic number* $f_g^+(G)$ of G [181]. Clearly, $f_g(G) \le f_g^+(G) \le g(G)$ in any graph G.

Theorem 2.23 ([173]). *For any positive integers a, b, c with $1 \le a \le b \le c-2$ or $4 \le a+2 \le b \le c$, there exists a connected graph G with $f_g(G) = a$, $f_g^+(G) = b$, and $g(G) = c$.*

Sketch of proof. In what follows, a graph G such that $f_g(G) = a$, $f_g^+(G) = b$, and $g(G) = c$ is called an (a,b,c)-graph. Check that the graphs G_1 and G_2 of Fig. 2.16 are a $(1,1,3)$-graph and a $(1,2,3)$-graph, respectively. For $k \ge 3$, let $G_k = (V,E)$ be the graph defined by $V = \{z\} \cup \{u_i, v_i\}_{i=1}^k \cup [\bigcup_{i=1}^{k-1} \{a_{ij}\}_{j=1}^7]$ and $E = \{zv_1, zu_1, v_1u_1, v_ku_k\} \cup \{u_iu_{i+1}, v_iv_{i+1}\}_{i=1}^{k-1} \cup [\bigcup_{i=1}^{k-1}\{a_{ij}a_{i(j+1)}\}_{j=1}^6] \cup \{a_{i2}a_{i7}, a_{i1}v_i, a_{i7}u_i\}_{i=2}^{k-1}$ (in Fig. 2.16 the graph G_3 is shown). Check that, for every $k \ge 3$, the graph G_k is a $(1, k, k+1)$-graph. We distinguish cases.

Case 1: $1 \le a \le b \le c-2$. If $a = b$, consider the graph H obtained from a copies $H_1, \ldots, H_a$ of G_1, by identifying y_{i-1} and x_i, where, for every $i \in \{1, \ldots, a\}$, $\mathrm{Ext}(H_i) = \{x_i, y_i\}$. Check that H is $(a, a, a+2)$-graph. Take $d = c - a - 1$ and consider d copies $B_1, \ldots, B_d$ of K_2, where, for every $i \in \{1, \ldots, d\}$, $V(B_i) = \{u_i, v_i\}$. Check that the graph G obtained from $H, B_1, \ldots, B_d$ by identifying $y_a, u_1, \ldots, u_d$, is a (a, a, c)-graph.

If $1 = a < b$, consider the graph G_b defined above. Take $d = c - b$ and check that the graph G obtained from $G_b, B_1, \ldots, B_d$ by identifying $z, u_1, \ldots, u_d$ is a $(1, b, c)$-graph.

If $2 = a < b$, take $d = c - b - 1$ and check that the graph G obtained from $G_{b-1}, G_1, B_1, \ldots, B_d$ by identifying $z, x, u_1, \ldots, u_d$, is a $(2, b, c)$-graph.

Finally, if $3 \le a < b$, take $k = b - a + 1$, $d = c - b - 1$ and check that the graph G obtained from $G_k, H, B_1, \ldots, B_d$ by identifying $z, x_1, u_1, \ldots, u_d$ is an (a, b, c)-graph.

Case 2: $4 \le a + 2 \le b \le c$. Consider the graph H_{a-2} obtained from G_2 and $a - 2$ copies $F_1, \ldots, F_{a-2}$ of G_1, by identifying the pairs $z = x_1, y_1 = x_2, \ldots, y_{a-3} = x_{a-2}$, where, for every $i \in \{1, \ldots, a-2\}$, $\text{Ext}(F_i) = \{x_i, y_i\}$ and $\text{Ext}(G_2) = \{z\}$. Take $d = c - b$ and check that the graph G obtained from $G_{b-a}, H_{a-2}, B_1, \ldots, B_d$ by identifying $z, y_{a-2}, u_1, \ldots, u_d$ is an (a, b, c)-graph.   □

2.6   Closed Geodomination

A geodetic set S of a graph G is a *closed geodetic set* if it can be written in canonical form as $S = \{v_1, v_2, \ldots, v_k\}$ such that $v_j \notin I_G[\{v_1, v_2, \ldots, v_{j-1}\}]$ for all $j \in \{2, \ldots, k\}$. The *closed geodetic number* of G, denoted by $cg(G)$, is the minimum cardinality of a closed geodetic set of G [21, 111].

Proposition 2.7 ([2]). *If G is a graph of order n and diameter d, then $g(G) \le cg(G) \le n - d + 1$. Moreover, if either $p \in \{1, 2, 3, n-1, n\}$ or G is an extreme geodesic graph, then $g(G) = cg(G) = p$.*

Sketch of proof. Take a pair of antipodal vertices $u_1, u_2 \in V(G)$. Notice that $|I[u_1, u_2]| \ge d + 1$. Complete the process to obtain a closed geodetic set $S = \{u_1, u_2, \ldots, u_k\}$. Hence, $n = |I[S]| \ge (d + 1) + (k - 2) = d + k - 1$, i.e., $g(G) \le cg(G) \le k \le n - d + 1$.

If G is an extreme geodesic graph, then $\text{Ext}(G)$ is a geodetic set, and it is also clearly a closed geodetic set. Thus, in this case, $g(G) = cg(G) = p$.

Notice that, $cg(G) = n$ if and only if $G \cong K_n$, since $g(G) \le cg(G) \le n - d + 1$.

Next, suppose that $cg(G) = n - 1$ and $g(G) \le n - 2$. If there exists a pair x, y of vertices such that either $d(x, y) \ge 3$ or $d(x, y) = 2$ and there are at least two $x - y$ geodesics, then $cg(G) \le n - 2$, a contradiction. Assume thus that $d = 2$ and that for every pair $x, y \in V(G)$ of nonadjacent vertices, $d(x, y) = 2$ and there is a unique $x - y$ geodesic. If $\text{rad}(G) = 1$, then prove, first, that $|\text{Cen}(G)| = 1$ and, second, that according to Theorem 2.5, $g(G) = n - 1$, a contradiction. If $\text{rad}(G) = 2$, prove that necessarily $cg(G) \le n - 2$, again a contradiction.

Finally, check that if S is a minimum geodetic set of cardinality at most 3, then it is also a closed geodetic set.   □

In particular, all graphs considered in Table 2.1 satisfy $g(G) = cg(G)$, except the family of complete bipartite graphs $K_{p,q}$ with $4 \le p \le q$. In this case, as was proved in [2], $cg(K_{p,q}) = p$. This result allowed the authors of this paper to obtain the following realization theorem.

Theorem 2.24 ([2]). *Given any triple of integers n, h, k with $4 \le h \le k - 1$ and $2k - h + 3 \le n$, there exists a graph G of order n such that $g(G) = h$ and $cg(G) = k$.*

Sketch of proof. Take $r = k - m + 3$ and $s = n - k$. Let $K_{r,s} = (U \cup V, E)$, where $U = \{u_i\}_{i=1}^r$ and $V = \{v_i\}_{i=1}^s$. Consider the graph G obtained from $K_{r,s}$ by adding a set $\{w_i\}_{i=1}^{m-3}$ of $m - 3$ leaves, all of them joined to vertex v_1. Notice that $\{w_i\}_{i=1}^{m-3} \cup \{u_1, u_r, v_s\}$ is a minimum geodetic set. Check that $\{w_i\}_{i=1}^{m-3} \cup \{u_i\}_{i=1}^r$ is minimum closed geodetic set.   □

2.7 Geodetic Domination

A vertex in a graph G *dominates* itself and its neighbors. A set of vertices S in a graph G is a *dominating set* if each vertex of G is dominated by some vertex of S. The domination number $\gamma(G)$ of G is the minimum cardinality of a dominating set of G. A dominating set S is called *geodetic dominating* if S is also a geodetic set of G. The *geodetic domination number* of G, denoted by $\gamma_g(G)$, is the minimum cardinality of a geodetic dominating set of G. Certainly, for every nontrivial connected graph G of order n, $2 \leq \max\{g(G), \gamma(G)\} \leq \gamma_g(G) \leq g(G) + \gamma(G)$, as the union of a geodetic set and a dominating set produces a geodetic dominating set. The *girth* $c(G)$ is the length of a shortest cycle contained in G.

Theorem 2.25 ([109]). *If G is a graph with $\delta(G) \geq 2$ and $c(G) \geq 6$, then $\gamma_g(G) = \gamma(G)$.*

Proof. Let D be a minimum dominating set of G. Suppose that $X = V(G) \setminus I[D] \neq \emptyset$ and take a vertex $x \in X$. Let $u \in N(x) \cap D$. Since $\delta(G) \geq 2$ and $c(G) \geq 4$, there is a vertex $v \in N(x) \setminus \{u\}$ and $uv \notin E(G)$. If $v \in D$, then $x \in I[u, v] \subseteq I[D]$, a contradiction. Thus $v \in V(G) \setminus D$, and there is a vertex $z \in N(v) \cap D$ as $c(G) \geq 6$, $d(u, z) \geq 3$. It follows that $x, v \in I[u, z] \subseteq I[D]$, a contradiction. Thus X is empty and D is a geodetic dominating set, as desired.   □

Theorem 2.26 ([97]). *Let G be a graph of order $n \geq 2$ and diameter* $\text{diam}(G) = d \geq 2$.

1. $\gamma_g(G) \leq n - \lfloor \frac{2d}{3} \rfloor$.
2. *If* $c(G) = c \geq 6$, *then* $\gamma_g(G) \leq n - \lfloor \frac{2c}{3} \rfloor$.

Moreover, both bounds are tight.

Proof.

1. Take nonnegative integers r, t such that $0 \leq r \leq 2$ and $d = 3t + r$. Take a pair u_0, u_d of antipodal vertices, i.e., s.t. $d(u_0, u_d) = d$. Let ρ be a $u_0 - u_d$ geodesic, where $V(\rho) = \{u_i\}_{i=0}^d$. Consider the set $A = \{u_0, u_3, \ldots, u_{3t}, u_d\}$. Check that the set $V(G) \setminus (V(\rho) \setminus A)$ is both geodetic and dominating. Observe that $t + 1 \leq |A| \leq t + 2$. Hence, $|V(G)| \setminus \gamma_g(G) \geq |(V(\rho) \setminus A| = \lfloor \frac{2d}{3} \rfloor = \lfloor \frac{2d}{3} \rfloor$.

   Finally, if $G \cong P_n$, then $\gamma_g(G) = \lceil \frac{n+2}{3} \rceil = n - \lfloor \frac{2(n-1)}{3} \rfloor = n - \lfloor \frac{2\text{diam}(P_n)}{3} \rfloor$.

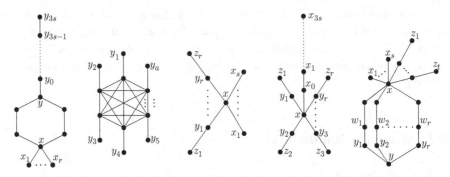

Fig. 2.17 $\gamma(G_1) = r+s+3$, $\gamma(G_2) = a$, $\gamma(G_3) = r+1$, $\gamma(G_4) = r+s+1$, $\gamma(G_1) = t+r+1$

2. Take nonnegative integers r,t such that $0 \le r \le 2$ and $c = c(G) = 3t+r$. Let $H \cong C_c$ be an induced cycle of length c, where $V(H) = \{w_i\}_{i=1}^c$. Take the set $A = \{u_1, u_4, \ldots, u_{3t-2}\}$ if $r = 0$ and $A = \{u_1, u_4, \ldots, u_{3t-2}, u_{3t+1}\}$ if $1 \le r \le 2$. Check that the set $V(G) \setminus (V(H) \setminus A)$ is both geodetic and dominating. Observe that $t \le |A| \le t+1$. Hence, $|V(G)| \setminus \gamma_g(G) \ge |(V(H) \setminus A| = \lfloor \frac{6t+2r}{3} \rfloor = \lfloor \frac{2c}{3} \rfloor$.

Finally, if $G \cong C_n$, then $\gamma_g(G) = \lceil \frac{n}{3} \rceil = n - \lfloor \frac{2n}{3} \rfloor = n - \lfloor \frac{2c(C_n)}{3} \rfloor$. □

Theorem 2.27 ([97]). *Let a,c,b be integers such that $a,b \ge 2$ and $\max\{a,b\} \le c \le a+b$. Then there is a connected graph G such that $\gamma(G) = a$, $g(G) = b$, and $\gamma_g(G) = c$.*

Sketch of proof. Take integers $a,b \ge 2$. We distinguish cases.

Case 1: $c = a + b$. Consider the graph G_1 shown in Fig. 2.17. Check that $\{x, y, y_2, \ldots, y_{3s-1}\}$ and $\{y_{3s}, x_1, \ldots, x_r\}$ are a minimum dominating set and a minimum geodetic set of G, respectively. Notice also that the union of these two sets produces a minimum geodetic dominating set. Hence, taking $r = b - 1$ and $s = a - 2$, we obtain the desired values.

Case 2: $a = b = c$. Consider the graph $G_2 \cong K_a \odot K_1$ shown in Fig. 2.17. Check that $\{y_i\}^a i = 1$ is a minimum geodetic dominating set.

Case 3: $a < b = c$. Consider the graph G_3 shown in Fig. 2.17. Check that $\{x, z_1, \ldots, z_r\}$ is a minimum dominating set and that $\{z_1, \ldots, z_r, x_1, \ldots, x_s\}$ is both a minimum geodetic set and a minimum geodetic dominating set. Hence, taking $r = a - 1$ and $s = b - a + 1$, we obtain the desired values.

Case 4: $b < a = c$. Consider the graph G_4 shown in Fig. 2.17. Check that $\{z_1, \ldots, z_r, x_0, x_3, \ldots, x_{3s}\}$ is both a minimum dominating set and a minimum geodetic dominating set and that $\{z_1, \ldots, z_r, x_{3s}\}$ is a minimum geodetic set. Hence, taking $r = b - 1$ and $s = a - b$, we obtain the desired values.

Case 5: $\max\{a,b\} < c < a+b$. Consider the graph G_5 shown in Fig. 2.17. Check that $\{x, z_1, \ldots, z_t, y_1, \ldots, y_r\}$ is a minimum dominating set, that $\{y, x_1, \ldots, x_s, z_1, \ldots, z_t\}$ is a minimum geodetic set, and that $\{y, x_1, \ldots, x_s, z_1, \ldots, z_t, w_1, \ldots, w_r\}$ is a minimum geodetic dominating set. Hence, taking $r = c - b$, $s = c - a$, and $t = a + b - c - 1$, we obtain the desired values. □

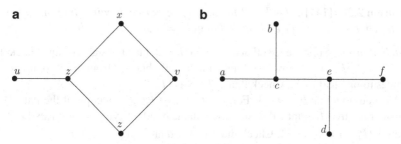

Fig. 2.18 (a) $g(G) = g_3(G) = |\{u,v\}|$, $g_2(G) = |\{u,x,y\}| = 3$. (b) $g(T) = g_2(G) = g_3(G) = 4$

2.8 *k*-Geodomination

Given a graph G of diameter d and an integer $2 \leq k \leq d$, a vertex v is said to be *k-geodominated* by a pair of vertices x and y if $d(x,y) = k$ and v lies on a shortest path between vertices x and y. A subset S of vertices in a graph $G = (V,E)$ is a *k-geodetic set* if each vertex in $V \setminus S$ is k-geodominated by some pair of vertices in S and the minimum cardinality of a k-geodetic set is the *k-geodetic number* of a graph $g_k(G)$ [147].

It is clear that, for every $k \in \{2, \ldots, d\}$, $2 \leq g(G) \leq g_k(G) \leq n - k + 1$, where $n \geq 2$ denotes the order of G.

Theorem 2.28 ([147]). *If G is a graph of diameter 2, then $g_2(G) = g(G)$.*

Proof. Let S be a minimum geodetic set of G. Since $\text{diam}(G) = 2$, it follows that $S \subsetneq V(G)$. Take an arbitrary vertex $v \notin S$. As S is a geodetic set, it follows that v is geodominated by a pair of vertices $x, y \in S$. Because $\text{diam}(G)$, it follows that $d(x,y)$. Thus S is a also two-geodetic set of G.   □

Certainly, the condition that the diameter of the graph is 2 is not necessary as the graph in Fig. 2.18a shows. Also, observe that the converse of this result is not true. For example, the tree T in Fig. 2.18b has diameter 3, and the set $\text{Ext}(T)$ of leaves of T is as minimum geodetic set as well as a minimum two-geodetic set.

Theorem 2.29 ([147]). *If $G = (V,E)$ is a graph of diameter $d \geq 3$ and $g(G) = 2$, then $g_k(G) = 2$ if and only if $k = d$.*

Sketch of proof. Let $S = \{x,y\}$ a minimum geodetic set of G. Then, x and y are antipodal, that is, $d(x,y) = d$. Thus S is a d-geodetic set and so $g_d(G) = 2$.

Conversely, if $k = 1$ or $k > d$, then V is the unique geodetic set of G, and so $g_k(G) = n > 2 = g(G)$, a contradiction. Assume next that $2 \leq k \leq d - 1$. Since $g(G) = 2$, every minimum geodetic set S contains antipodal vertices and so S is d-geodetic. As $k < d$, it follows that S is not k-geodetic. Thus no two subset of V is a k-geodetic set. Therefore, $g_k(G) > 2 = g(G)$, again a contradiction.   □

Theorem 2.30 ([147]). *Let $k \geq 2$ be an integer. For each pair a,b of integers with $2 \leq a \leq b$, there exists a tree T such that $g(T) = a$ and $g_k(T) = b$.*

Sketch of proof. Suppose first that $a = b$. For $k = 2$, take the star $K_{1,a}$ and check that $g(K_{1,a}) = g_2(K_{1,a}) = a$. For $k \geq 3$, take the tree T obtained from P_k by joining $a - 1$ vertices to one leaf of P_k. Check that $g(T) = g_k(T) = a$.

Assume now that $b = a + 1$. For $1 \leq i \leq a$, let P_{k+1}^i be a copy of the path P_{k+1}. Consider the tree T obtained from these a paths by identifying the vertices $\{x_{i1}\}_{i=1}^a$, where $V(P_{k+1}^i) = \{x_{ij}\}_{j=1}^{k+1}$. Check that $g(T) = a$ and $g_k(T) = a + 1$.

Finally, suppose that $b = a + j$, with $j \geq 2$. Consider the path P_{jk}. It is routine to check that $g_k(P_{jk}) = j + 2$. Hence, if $a = 2$, P_{jk} fulfills the desired property. So, we may assume that $a \geq 3$. Let T be the tree obtained from P_{jk} by joining $a - 2$ vertices to one support vertex of this path. Check that $g(T) = a$ and $g_k(T) = j + a = b$. □

Conjecture 2.5. For any integers a,b,n,k,d with $2 \leq k \leq d$ and $2 \leq a \leq b \leq n-k+1$, there exists a graph G of order n and diameter d such that $g(G) = a$ and $g_k(G) = b$.

Conjecture 2.6. For any integers $2 \leq a,b,h,k$ there exists a graph G such that $g_h(G) = a$ and $g_k(G) = b$.

Theorem 2.31 ([13]). *If $G = (V,E)$ is a graph of order n, diameter d, and maximum degree Δ, then, for every $k \in \{2,\ldots,d\}$, $g_k(G) \geq \lceil \frac{2n}{\Delta(\Delta-1)^{k-1}(k-1)+2} \rceil$.*

Proof. Let S be a k-geodetic set of G. We know that every vertex not in S lies on a geodesic of length k joining to vertices of S. For every vertex u, the number of length k beginning in u is bounded above by $\Delta(\Delta - 1)^{k-1}$. If we consider all vertices in S, we have that the number of paths of length k beginning and ending in S is at most $\frac{|S|\Delta(\Delta-1)^{k-1}}{2}$ and the number of vertices of $V(G) \setminus S$ which can lie on those paths is bounded above by $\frac{|S|\Delta(\Delta-1)^{k-1}(k-1)}{2}$. Therefore, $n - |S| \leq \frac{|S|\Delta(\Delta-1)^{k-1}(k-1)}{2}$. □

2.9 Edge Geodomination

A set of vertices U of a graph G is said to be *edge geodetic* if the union of all edges belonging to all geodesics joining pairs of vertices of U is the whole edge set $E(G)$ of G [8]. The *edge geodetic number* $g_e(G)$ of G is defined as the minimum cardinality of an edge geodetic set of G.

Clearly, if G is a nontrivial graph of order n and diameter d, then $2 \leq g(G) \leq g_e(G) \leq \min\{g(G)+n-2, n-d+1\}$. The equality $g(G) = g_e(G)$ holds, among others, for trees, cycles, and complete graphs. The equality $g_e(G) = g(G) + n - 2$ holds, for example, for the graph $K_n - e$, obtained from K_n by deleting one edge e. Finally, the equality $g_e(G) = n - d + 1$ holds, for example, for paths, stars, and complete graphs.

A collection $\mathscr{S}$ of subsets of $[p]$ is said to *separate pairs in* $[p]$ if, for any two distinct $i, j \in [p]$, there exists a pair of disjoint sets $A, A' \in \mathscr{S}$ such that $i \in A$ and $j \in A'$.

Lemma 2.2 ([8]). *The minimum cardinality of a collection $\mathscr{S}$ of subsets $[p]$ that separates pairs in $[p]$ is $\lceil 3 \log_3 p \rceil$.*

Theorem 2.32 ([7,8]). *For any connected graph G, $\lceil 3 \log_3 \omega(G) \rceil \le g_e(G)$. Moreover, this bound is tight.*

Sketch of proof. Take a maximum clique $\Omega_p \cong K_p$ of G, where $p = \omega(G)$ and $V(K_p) = [p]$. Take a minimum edge geodetic set U of G. Given a vertex $u \in U$ and a geodesic γ starting at u such that $E(\gamma) \cap E(\Omega_p) \ne \emptyset$, let u_γ denote the vertex in $V(\gamma) \cap V(\Omega_p)$ closer to u.

Consider the set $\mathscr{S} = \{A_u\}_{u \in U}$, where $A_u = \{u_\gamma : \gamma$ is a geodesic with endvertex $u\}$. Check that $\mathscr{S}$ separates pairs in $[p]$. Hence, $g_e(G) = |U| = |\mathscr{S}| \ge \lceil 3 \log_3 p \rceil$.

Finally, to prove the sharpness, given a positive integer p, take the complete graph K_p and a set $S = \{u_i\}_{i=1}^q$ of $q = \lceil 3 \log_3 p \rceil$ new vertices. Take a collection $\mathscr{S} = \{S_i\}_{i=1}^q$ of $q = \lceil 3 \log_3 p \rceil$ subsets of $[p] = V(K_p)$ that separates pairs in $[p]$. Consider that graph G obtained from K_p and S by joining, for every $i \in \{1, \dots, q\}$, vertex u_i to all vertices of S_i. Check that S is a minimum edge geodetic set of G. □

Theorem 2.33 ([166]). *For each pair a, b of integers with $2 \le a \le b$, there exists a graph G such that $g(G) = a$ and $g_e(G) = b$.*

Sketch of proof. If $a = b$, take the star $K_{1,a}$. If $2 = a < b$, take the graph $K_1 \vee K_{2,b-2}$. Finally, if $2 < a < b$, take the graph G obtained from $H \cong K_1 \vee K_{2,b-2}$ by adding $a - 2$ leaves, all of them joined to the universal vertex of H. □

Theorem 2.34 ([118]). *The boundary $\partial(G)$ of every graph is an edge geodetic set. Moreover, if G is bipartite, then for every $u \in V(G)$, $\{u\} \cup \partial(u)$ is an edge geodetic set.*

Proof. Let $e = xy \in E(G)$. Take a maximal geodesic ρ containing e. This means that if a, b are the endpoints vertices of ρ, then $b \in \partial(a)$ and $a \in \partial(b)$. Thus e is in a geodesic with endpoints in $\partial(G)$.

Let $u \in V(G)$ and $e = (x, y) \in E(G)$. Because G is a bipartite graph, $d(x, u) \ne d(y, u)$. Suppose that $d(x, u) < d(y, u)$. Take a $u - x$ geodesic ρ. Then, the path ρ' obtained from ρ by joining vertex y with the edge e is a $u - y$ geodesic that can be extended to a maximal $u - z$ geodesic. Then $z \in \partial(u)$ and hence e is in a geodesic with endpoints in $\{u\} \cup \partial(u)$. □

Further results involving geodominating invariants related to geodesic convexity in graphs can be found in [2, 7, 8, 13, 49, 52, 53, 58, 60, 77, 79, 97, 109, 111, 125, 126, 137, 146, 147, 166, 171, 173, 176, 177, 181].

2.10 Classical Parameters

A subset $A \subset V$ of a graph $G = (V, E)$ is called *Helly independent*[2] if $\bigcap_{a \in A}[A - a] = \emptyset$. The *Helly number* $H(G)$ of a graph G is the maximum cardinality of a Helly independent set. In other words, $H(G)$ is the smallest integer such that every family of convex sets with an empty intersection contains a subfamily of at most $H(G)$ members with an empty intersection.

Given a subset $A \subset V$ of a graph $G = (V, E)$, a partition $\{A_1, A_2\}$ of A is called a *Radon partition* if $[A_1] \cap [A_2] \neq \emptyset$. A subset $A \subset V$ of a graph $G = (V, E)$ is called *Radon dependent* if it has a Radon partition, and otherwise it is said to be *Radon independent*. The *Radon number*[3] $R(G)$ of a graph $G = (V, E)$ is the maximum cardinality of a Radon independent set.

A subset $A \subset V$ of a graph $G = (V, E)$ is called *redundant* when $[A] = \bigcup_{a \in A}[A - a]$, and otherwise it is said to be *irredundant*. The *Carathéodory number* $C(G)$ of a graph G is the maximum cardinality of an irredundant set.

Theorem 2.35 ([138]). *Let $G = (V, E)$ be a connected graph of order $n \geq 2$.*

1. *Every subset of a Helly independent set is Helly independent.*
2. *Every subset of a Radon independent set is Radon independent.*
3. $2 \leq \omega(G) \leq H(G)$.
4. $H(G) \leq R(G)$.

Proof.

1. Take $B \subset A \subseteq V$. Suppose that B is Helly dependent. Observe that $\cap_{b \in B}[B - b] \subseteq \cap_{b \in B}[A - b]$. Note also that, for every $a \in A - B$, $\cap_{b \in B}[B - b] \subseteq [B] \subseteq [A - a]$. Hence, $\emptyset \neq \cap_{b \in B}[B - b] \subseteq (\cap_{b \in B}[A - b]) \cap (\cap_{a \in A - B}[A - a]) = \cap_{a \in A}[A - a]$.

Table 2.4 Classical parameters of some basic graph families

G[a]	P_n[b]	C_n[c]	T_n[d]	K_n	$K_{p,q}$[e]	Q_n[f]
$\omega(G)$	2	2	2	n	2	2
$H(G)$	2	3	2	n	2	2
$R(G)$	2	3	3	n	2	$\lceil \log_2(n+1) \rceil + 1$
$C(G)$	2	2	2	1	2	n

[a] $G \ncong K_1$
[b] $2 < n$
[c] $4 < n$
[d] $\Delta(T_n) \geq 3$
[e] $2 \leq p \leq q$
[f] $2 < n$

---

[2] As in the remaining sections and chapters, unless otherwise stated, all terms, invariants and results are referred to the geodesic convexity.

[3] Some authors [90, 132] define the Radon number to be one unit larger, i.e., as the smaller value r such that each set with at least r points admits a Radon partition.

2. Take $B \subset A \subseteq V$. Suppose that B is Radon dependent. Let $\{B_1, B_2\}$ be a Radon partition of B. Consider the partition of A, $\{B_1, A_2\}$, where $A_2 = B_2 \cup (A - B)$. Then, as $\emptyset \neq [B_1] \cap [B_2] \subseteq [B_1] \cap [A_2]$, it follows that $\{B_1, A_2\}$ is a Radon partition of A.

3. Let $A \subseteq V$ such that $G[A] \cong K_h$. Then, for every $a \in A$, $G[A - a] \cong K_{h-1}$, which means that $\cap_{a \in A}[A - a] = \emptyset$.

4. Let $A \subseteq V$ be Radon dependent. Consider a Radon partition $\{A_1, A_2\}$ of A and a vertex $p \in [A_1] \cap [A_2]$. For each $a \in A$, we have either $A_1 \subseteq A - a$ or $A_2 \subseteq A - a$. Hence, $p \in [A - a]$ for every $a \in A$, which means that A is Helly dependent. $\quad\square$

Remark 2.1. Consider the Euclidean convexity, derived from the Euclidean distance, in R^n. Then, the Helly number, the Radon number, and the Carathéodory number are all equal to $n + 1$ [175].

Let $W \subset V(G)$ be a set of vertices of a connected graph G such that $G[W]$ is connected. Take the subgraph $G - W$ and a new vertex x_W, and, finally, join this vertex to every vertex in $y \in (V(G) \setminus W) \cap N_G(W)$. This operation is called a *contraction of a connected subgraph* of G. A *contraction* of G is every graph that can be obtained by a sequence of contractions of connected subgraphs. The *Hadwiger number* $\eta(G)$ of G is the number of vertices in the largest clique that can be formed as a contraction of G.

Theorem 2.36 ([93]). *For every connected graph G, $H(G) \leq \eta(G)$, $R(G) \leq 2\eta(G)$. Moreover, the first bound is tight.*

According to Duchet's research [90, 91], the geodesic convexity is in the following sense universal with respect to Carathéodory, Helly, and Radon parameters: Given any finite convexity space $(X, \mathscr{C})$, there exists a finite graph G such that the Carathéodory, Helly, and Radon numbers of the geodesic convexity in G coincide with those of $\mathscr{C}$.

Theorem 2.37 ([89, 129]). *Let G be a non-complete connected graph. Let $H_m(G)$, $R_m(G)$, and $C_m(G)$ denote the Helly, Radon, and Carathéodory numbers of G with respect to the monophonic convexity. Then,*

1. $H_m(G) = \omega(G)$.
2. $R_m(G) = \omega(G)$ if $\omega(G) \geq 3$ and $2 \leq R_m(G) \leq 3$ if $\omega(G) = 2$.
3. $C_m(G) = 2$.

A *distance-hereditary graph* is a graph in which every chordless path is a geodesic [120].

Corollary 2.3. *If G is a non-complete distance-hereditary graph, then $C(G) = 2$, $H(G) = \omega(G)$, and $R(G) = \omega(G)$ if $\omega(G) \geq 3$ ($2 \leq R(G) \leq 3$ if $\omega(G) = 2$).*

In the case of the geodesic convexity of a graph G, the Helly number $H(G)$ may well exceed the clique number $\omega(G)$, the five-cycle C_5 being the smallest example.

A graph is *weakly modular* if, for any triple $\{x,y,z\}$ of vertices such that $I[x,y] \cap I[x,z] = \{x\}$, $I[y,x] \cap I[y,z] = \{y\}$, and $I[z,x] \cap I[z,y] = \{z\}$, all vertices on every $y-z$ geodesic have the same distance to x.

A graph is called *constructible* if there exists a well-ordering $\leq$ of its vertices such that, for every vertex x which is not the smallest element, there is a vertex $y < x$ which is adjacent to x and to every neighbor z of x with $z < x$.

Note that there exists no relation between constructible graphs and weakly modular graphs, as is shown by the following two examples. Let C_4 be a cycle of length four. Then C_4 is clearly weakly modular but not constructible. Now let x and y be two new vertices, and let G be the graph obtained by joining x to all vertices of C_4 and y to two adjacent vertices of C_4. Then G is constructible but not weakly modular.

Theorem 2.38 ([9, 162]). *Every weakly modular (resp. constructible) graph G satisfies* $\omega(G) = H(G)$.

Further results on classical convexity invariants directly or indirectly related to geodesic convexity in graphs can be found in [9, 12, 40, 43, 87, 89–93, 128, 129, 132, 137, 149, 159, 162, 167, 168, 175].

Chapter 3
Graph Operations

3.1 Cartesian Product

The *Cartesian product* of graphs $G_1 = (V_1, E_1)$ and $G_2 = (V_2, E_2)$ is the graph $G_1 \square G_2$ on vertex set $V_1 \times V_2$ in which vertices (x_1, x_2) and (y_1, y_2) are adjacent if and only if either $x_1 = x_2$ and $y_1 y_2 \in E_2$ or $y_1 = y_2$ and $x_1 x_2 \in E_1$.

The map $p_{G_1} : V(G_1) \times V(G_2) \to V(G_1)$, $p_{G_1}((g_1, g_2)) = g_1$, is the *projection* onto G_1 and $p_{G_2} : V(G_1) \times V(G_2) \to V(G_2)$, $p_{G_2}((g_1, g_2)) = g_2$, the projection onto G_2.

Theorem 3.1 ([17, 108, 130]). *Let G, H be connected graphs. If $S \subseteq V(G \square H)$, $(g, h), (g', h') \in V(G \square H)$ and ρ is a $(g, h) - (g', h')$ geodesic in $G \square H$, then*

1. $d_{G \square H}((g, h), (g', h')) = d_G(g, g') + d_H(h, h')$.
2. $p_G(P)$ *induces a* $g - g'$ *geodesic in* G *and* $p_H(P)$ *induces an* $h - h'$ *geodesic in* H.
3. $I_{G \square H}[(g, h), (g', h')] = I_G[g, g'] \times I_H[h, h']$.
4. $I_{G \square H}[S] \subseteq I_G[p_G(S)] \times I_H[p_H(S)]$.
5. $[S] = [p_G(S)] \times [p_H(S)]$.

Proof.

1. $d_{G \square H}((g, h), (g', h')) = d_{G \square H}((g, h), (g', h)) + d_{G \square H}((g', h), (g', h')) = d_G(g, g') + d_H(h, h')$.
2. If $A = V(\rho) = \{(g, h)\} \cup \{(g_i, h_i)\}_{i=1}^{h-1} \cup \{(g', h')\}$, then $p_G(A) = \{g\} \cup \{g_i\}_{i=1}^{h-1} \cup \{g'\}$ induces a $g - g'$ path in G and $p_H(A) = \{h\} \cup \{h_i\}_{i=1}^{h-1} \cup \{h'\}$ induces a $h - h'$ path in H, and according to the previous item, both paths must be geodesics.
3. Consider any $x = (x_1, x_2) \in I_{G \square H}[a, b]$, along an $a - b$ geodesic, where $a = (a_1, a_2)$, $b = (b_1, b_2)$. Then since $d_{G \square H}(a, b) = d_G(a_1, b_1) + d_H(a_2, b_2)$, it must be for $i = 1, 2$ that x_i is on an $a_i - b_i$ geodesic, so $x \in I_G[a_1, b_1] \times I_H[a_2, b_2]$.

   Conversely, consider any $x = (x_1, x_2) \in I_G[a_1, b_1] \times I_H[a_2, b_2]$. Then there exists an $a_1 - b_1$ geodesic which follows a path P from a_1 to x_1 and from there on to b_1 via a path Q and likewise an $a_2 - b_2$ geodesic which follows a path R through x_2. Then the path from $a = (a_1, a_2)$ to (x_1, a_2) (formed as in P while

I.M. Pelayo, *Geodesic Convexity in Graphs*, SpringerBriefs in Mathematics, DOI 10.1007/978-1-4614-8699-2_3, © Ignacio M. Pelayo 2013

holding the second entry a_2 fixed) followed by the path from there to (x_1,x_2) and then on to x_1,b_2 (formed as in R while holding the first entry x_1 fixed) followed by the path from there to (b_1,b_2) (formed as in Q while holding the second entry b_1 fixed) yields an $a-b$ geodesic containing x. Therefore item 3 holds.

4. $I[S] = \bigcup_{a,b \in S} I[a,b] = \bigcup_{a,b \in S} I[a_1,b_1] \times I[a_2,b_2] \subseteq (\bigcup_{a_1,b_1 \in p_G(S)} I[a_1,b_1]) \times (\bigcup_{a_2,b_2 \in p_H(S)} I[a_2,b_2]) = I[p_G(S)] \times I[p_G(S)]$.

5. Consider any convex set S of $G \square H$. We show that $p_G([S])$ is a convex set of G as follows. Suppose for contradiction that for some $x,x' \in p_G([S])$ there exists an $x-x'$ geodesic in G containing a vertex $z \notin p_G([S])$. Then there exist $(x,y),(x',y') \in [S]$. The distance from (x,y) to (x',y') is $d_G(x,x') + d_H(y,y')$, so there exists an $(x,y)-(x',y')$ geodesic that goes through (z,y) along the way to (x',y) before continuing on to (x',y'). Since $[S]$ is convex, we have that $(z,y) \in [S]$, so $z \in p_G([S])$, a contradiction. Therefore, $p_G([S])$ is a convex set of G, and hence $[p_G(S)] \subseteq p_G([S])$. By symmetric argument, $p_H([S])$ is a convex set of H, and hence $[p_H(S)] \subseteq p_H([S])$.

   Next we show that $[p_G(S)] \times [p_H(S)] \subseteq [S]$. Consider any $(x,y) \in [p_G(S)] \times [p_H(S)]$. There exist x',y' for which $(x,y') \in [S]$ and $(x',y) \in [S]$, since $[p_G(S)] \subseteq p_G([S])$ and $[p_H(S)] \subseteq p_H([S])$. There is an $(x,y')-(x',y)$ geodesic that passes through (x,y), so $(x,y) \in [S]$. Therefore $[p_G(S)] \times [p_H(S)] \subseteq [S]$. To get the equality, it is enough to notice that $[p_G(S)] \times [p_H(S)]$ is a convex set of $G \square H$ containing S. $\qquad\square$

Corollary 3.1 ([17,35,130]). *Let G,H be connected graphs. If $W \subseteq V(G \square H)$, $S \subseteq V(G)$, and $T \subseteq V(H)$, then*

- W *is a convex set of $G \square H$ if and only if $p_G(W)$ is a convex set of G, $p_H(W)$ is a convex set of H, and $W = p_G(W) \times p_H(W)$.*
- *If W is a geodetic set of $G \square H$, then $p_G(W)$ is a geodetic set of G and $p_H(W)$ is a geodetic set of H.*
- $S \times T$ *is a geodetic set of $G \square H$ if and only if S is a geodetic set of G and T is a geodetic set of H.*

Corollary 3.2 ([35]). *Let G,H be connected graphs.*
Then, $\text{con}(G \square H) = \max\{|V(G)| \cdot \text{con}(H), |V(H)| \cdot \text{con}(G)\}$.

Theorem 3.2 ([17,130]). *Let G,H be connected graphs with $g(G) = p \geq g(H) = q \geq 2$. Then, $p \leq g(G \square H) \leq pq - q$.*

Proof. The lower bound is directly derived from the second item of Corollary 3.1. To show the upper bound, let $S = \{g_i\}_{i=1}^{p}$ and $T = \{h_i\}_{i=1}^{q}$ be geodetic sets of G and H, respectively. We claim that the set $U = (S \times T) \setminus \{(g_i,h_i)\}_{i=1}^{q}$ is a geodetic set of $G \square H$. Let (g,h) be an arbitrary vertex of $G \square H$. Then there exist indices i,i' such that $g \in I_G[g_i,g_{i'}]$, and there are indices j,j' such that $h \in I_G[h_j,h_{j'}]$. Since $2 \leq q \leq P$, we may assume that $i \neq i'$ and $j \neq j'$. Take the set $B = \{(g_i,h_j),(g_i,h_{j'}),(g_{i'},h_j),(g_{i'},h_{j'})\}$. Suppose that one of the vertices from B is not in U. We may w.l.o.g assume $(g_i,h_i) \notin U$. This means that $i = j$, and hence $i' \neq j$ and $i \neq j'$. Then we infer, according to item 3 of Theorem 3.1, that $(g,h) \in I_{G \square H}[(g_i, h_{j'}), (g_{i'}, h_j)]$. Otherwise, if all vertices of B and in U, then $(g,h) \in I_{G \square H}[(g_i, h_j), (g_{i'}, h_{j'})]$.   □

Theorem 3.3 ([130]). *The bounds of Theorem 3.2 are tight.*

Sketch of proof. To prove the sharpness of the lower bound take $G \cong K_p$ and $H \cong K_q$ with $V(G) = \{u_i\}_{i=1}^p$ and $V(G) = \{v_i\}_{i=1}^q$ and check that $\{(u_i, v_i)\}_{i=1}^q \cup \{(u_{q+i}, v_q)\}_{i=1}^{p-q}$ is a geodetic set of $G \square H$. Hence, $g(G \square H) \leq |S| = p = \max\{g(G), g(H)\} \leq g(G \square H)$, so equality holds.

To show the sharpness of the upper bound, consider the graph D_p^t designed as follows:

- Take p vertices $x_1, \ldots, x_p$.
- Take $\binom{p}{2}$ groups of vertices $W_{i,j}$ for $i, j \in \{1, \ldots, p\}$, $i < j$, where the $W_{i,j}$'s are pairwise disjoint and also disjoint from $\{x_1, \ldots, x_p\}$ and each $W_{i,j}$ consists of t vertices.
- For each pair $i, j \in \{1, \ldots, p\}$, $i < j$, join each of the t vertices in $W_{i,j}$ to both x_i and x_j.
- Finally, add a new vertex z and join it to all the other vertices.

Check that if $t > pq - p$, $G \cong D_p^t$, and $H \cong K_q$, then $g(G) = p$ and $g(G \square H) = pq - p$.   □

Let G be a graph and let $S = \{x_1, \ldots, x_k\}$ be a geodetic set of G. We say that S is a *linear geodetic* set if for any $x \in V(G)$ there exists an index i, $1 \leq i < k$, such that $x \in I[x_i, x_{i+1}]$. Complete graphs, complete bipartite graphs, cycles, and graphs G with $g(G) = 2$ are basic instances of graphs admitting linear minimum geodetic sets.

Theorem 3.4 ([17]). *Let G, H be nontrivial connected graphs with $g(G) = p$ and $g(H) = q$. If both G and H contain linear minimum geodetic sets, then $g(G \square H) \leq \lfloor \frac{pq}{2} \rfloor$.*

Proof. Let $S = \{g_i\}_{i=1}^p$ and $T = \{h_i\}_{i=1}^q$ be linear geodetic sets of G and H, respectively. We claim that the set $U = \{(g_i, h_j) : 1 \leq i \leq p, \leq j \leq q, i + j \text{ odd}\}$ is a geodetic set of $G \square H$. Let (g, h) be an arbitrary vertex of $G \square H$. Since S is linear, there exists an index i, $1 \leq i < p$, such that $g \in I_G[g_i, g_{i+1}]$, and because T is linear, there exists an index j, $1 \leq j < q$, such that $h \in I_H[h_j, h_{j+1}]$.

Suppose that $i + j$ is odd. Then $(i+1) + (j+1)$ is odd as well, which means that $\{(g_i, h_j), (g_{i+1}, h_{j+1})\} \subseteq U$. Hence, $(g,h) \in I_{G \square H}[(g_i, h_j), (g_{i+1}, h_{j+1})] \subseteq I_{G \square H}[U]$. Assume next that $i + j$ is even. Then, both $i + (j+1)$ and $(i+1) + j$ are odd and $\{(g_i, h_{j+1}), (g_{i+1}, h_j)\} \subseteq U$. Hence, $(g,h) \in I_{G \square H}[(g_i, h_{j+1}), (g_{i+1}, h_j)] \subseteq I_{G \square H}[U]$. To conclude the proof notice that $|U| = \lfloor \frac{pq}{2} \rfloor$.   □

Let G be a graph and let S be a geodetic set of G. We say that S is a *complete geodetic* set if for any $x \in V(G) \setminus S$ and for any $v, w \in S$, $u \in I[v, w]$. Clearly every complete geodetic set is also a linear geodetic set. Complete graphs, stars,

Table 3.1 Geodetic number
of $G \Box H$, for some basics
families [17, 130, 176, 178]

G/H	P_n	K_n	C_{2l}	C_{2l+1}
P_m	2	n	2	3
K_m	m	n	m	$m+1$
C_{2h}	2	n	2	3
C_{2h+1}	3	$n+1$	3	4
$2 \leq 2 \leq m \leq n, 2 \leq h, l$				

and graphs with geodetic number 2 are basic instances of graphs admitting linear minimum geodetic sets. Odd cycles are examples of graphs that admit a linear minimum geodetic set but not a complete minimum geodetic set. In general, trees need not have any complete minimum geodetic set.

Theorem 3.5 ([17]). *Let G, H be nontrivial connected graphs with $g(G) = p$ and $g(H) = q$. If $q \leq p$ and both graphs have a complete minimum geodetic set, then $g(G \Box H) = p$.*

Proof. Let $S = \{g_i\}_{i=1}^{p}$ and $T = \{h_i\}_{i=1}^{q}$ be complete geodetic sets of G and H, respectively. We claim that the set $U = \{(g_i, h_i)\}_{i=1}^{q} \cup \{(g_{q+i}, h_q)\}_{i=1}^{p-q}$ is a geodetic set of $G \Box H$. Let (g, h) be an arbitrary vertex of $G \Box H$. We distinguish cases.

Case 1: $g \in S$ and $h \in T$. Then there are indices i, j such that $g = g_i$ and $h = h_j$. We may assume $i \neq j$, otherwise $(g, h) \in U$. Note that $(g, h) \in I_{G \Box H}[(g_j, h_j), (g_i, h_\sigma)] \subseteq I_{G \Box H}[U]$, where $\sigma = i$ if $i \leq q$ and $\sigma = q$ if $i > q$.

Case 2: $g \in S$ and $h \notin T$. Then $g = g_i$ for some $i \in \{1, \ldots, p\}$. Let σ be defined as in the previous case, and let $j \leq q$ be an integer different from σ. Since $h \in I_H[h_j, h_\sigma]$, we derive $(g, h) \in I_{G \Box H}[(g_j, h_j), (g_i, h_\sigma)] \subseteq I_{G \Box H}[U]$.

Case 3: $g \notin S$ and $h \in T$. Then $h = h_j$ for some $j \in \{1, \ldots, q\}$. Let $i \leq q$ be an integer different from j. Then clearly $(g, h) \in I_{G \Box H}[(g_i, h_i), (g_j, h_j)] \subseteq I_{G \Box H}[U]$.

Case 4: $g \notin S$ and $h \notin T$. In this case, we have $(g, h) \in I_{G \Box H}[(g_1, h_1), (g_2, h_2)] \subseteq I_{G \Box H}[U]$.

Hence, U is a geodetic set, and since $|U| = p$, we conclude, according to Theorem 3.2, that U is a minimum geodetic set of $G \Box H$, i.e., $g(G \Box H) = p$. $\Box$

As a straightforward consequence of the previous theorems, along with some extra arguments when one of the factors is an odd cycle, a number of results on the geodetic number of Cartesian product graphs is displayed in Table 3.1.

Theorem 3.6 ([17]). *Let T_1, T_2 be trees. Then, $g(T_1 \Box T_2) = \max\{\mathrm{Ext}(T_1), \mathrm{Ext}(T_2)\}$.*

Proof. Suppose $p \geq g(T_1) \geq g(T_2) \geq q$. Notice that the unique geodetic set of a tree T is the set $\mathrm{Ext}(T)$ of its leaves. If $L_1 = \mathrm{Ext}(T_1) = \{x_i\}_{i=1}^{p}$, $L_2 = \mathrm{Ext}(T_2) = \{y_i\}_{i=1}^{q}$, and $f : L_1 \mapsto L_2$ is an arbitrary subjective mapping, we claim that the set $U = \{(x_i, f(x_i))\}_{i=1}^{p}$ is a geodetic set of $T_1 \Box T_2$. Let (g, h) be an arbitrary vertex of $T_1 \Box T_2$. Clearly there exist $x_i, x_j \in L_1$ such that $g \in I_{T_1}[x_i, x_j]$. If $h \in I_{T_2}[f(x_i), f(x_j)]$, then $(g, h) \in I_{G \Box H}[(x_i, f(x_i)), (x_j, f(x_j))] \subseteq I_{G \Box H}[U]$, as desired. Suppose next that $h \notin I_{T_2}[f(x_i), f(x_j)]$. Observe that there exists a leaf y in T_2 such that $h \in I_{T_2}[f(x_i), y]$.

Notice also that $h \in I_{T_2}[f(x_j),y]$. Choose $x \in f^{-1}(y)$. Since $g \in I_{T_1}[x_i,x_j]$, we conclude that either $g \in I_{T_1}[x_i,x]$ or $g \in I_{T_1}[x,x_j]$. Hence, in any case, we obtain that $(g,h) \in I_{G\square H}[U]$.   □

Theorem 3.7 ([29, 130]). *Let G,H be connected graphs.*
Then, $h(G\square H) = \max\{h(G),h(H)\}$.

Proof. Let $A = \{a_1,\ldots,a_p\}$ be a minimum g-hull set of G and let $B = \{b_1,\ldots,b_q\}$ be a minimum g-hull set of H, where without loss of generality we can assume that $p \geq q$. Consider any $S \subseteq V(G \times H)$. By Theorem 3.1, $[S] = [p_G(S)] \times [p_G(S)]$, so S is a g-hull set of $G \times H$ if and only if $p_G(S)$ is a g-hull set of G and $p_G(S)$ is a g-hull set of H. The choice $S = \{(a_i,b_i)\}_{i=1}^{q} \cup \{(a_{q+i},b_q)\}_{i=1}^{p-q}$ satisfies this requirement, with $|S| = p$, so $h(G) \leq p$. Also, $h(G \times H) \geq p$, since S must have at least $h(G)$ many elements for its projection $p_G(S)$ to have at least $h(G)$ elements. Thus $h(G \times H) = p$, completing the proof.   □

3.2   Strong Product

The *strong product* of graphs $G_1 = (V_1,E_1)$ and $G_2 = (V_2,E_2)$ is the graph $G_1 \boxtimes G_2$ on vertex set $V_1 \times V_2$ in which vertices (x_1,x_2) and (y_1,y_2) are adjacent whenever (1) $x_1 = x_2$ and $y_1y_2 \in E_2$, or (2) $y_1 = y_2$ and $x_1x_2 \in E_1$, or (3) $x_1x_2 \in E_1$ and $y_1y_2 \in E_2$.

Next, some basic properties of this graph operation are shown. The proofs are a direct consequence of the definition and thus omitted.

Theorem 3.8 ([27, 108]). *Let G,H be graphs.*

1. *$G \boxtimes H$ is connected if and only if both G and H are connected.*
2. *$G \boxtimes H$ is complete if and only if both G and H are complete.*
3. *If G is connected, then $d_{G\boxtimes H}((g,h),(g',h')) = \max\{d_G(g,g'),d_H(h,h')\}$.*
4. *$\text{diam}(G \boxtimes H) = \max\{\text{diam}(G),\text{diam}(H))\}$.*
5. *$\omega(G \boxtimes H) = \omega(G)\omega(H)$.*
6. *$\text{Ext}(G \boxtimes H) = \text{Ext}(G) \times \text{Ext}(H)$.*
7. *$G \boxtimes H$ is an extreme geodesic graph if and only if both G and H are extreme geodesic graphs.*
8. *If both G and H are extreme geodesic graphs, then $h(G \boxtimes H) = g(G \boxtimes H) = g(G)g(H) = h(G)h(H)$.*

As a direct consequence of the last property, the results shown in Table 3.2 are obtained.

Table 3.2 Hull and geodetic number of $G \boxtimes H$, for some basics families [27]

G/H	P_n	T_n^k	K_n
P_m	4	$2k$	$2n$
T_m^h	$2h$	hk	hn
K_m	$2m$	mk	mn

Table 3.3 Hull and geodetic number of $G \boxtimes H$, for some basics families [27]

$G \boxtimes C_n$	$h(G \boxtimes C_n)$	$g(G \boxtimes C_n)$
$P_m \boxtimes C_n$	$\begin{cases} 3, & \text{if } n = 2r+1 \text{ odd and } m < r+2 \\ 2, & \text{otherwise} \end{cases}$	$\begin{cases} 4, & \text{if } n \text{ is even} \\ 5-6, & \text{if } n \text{ is odd} \end{cases}$
$K_m \boxtimes C_n$	$\begin{cases} 2, & \text{if } n \text{ is even} \\ 3, & \text{if } n \text{ is odd} \end{cases}$	$\begin{cases} 4, & \text{if } n \text{ is even} \\ 5, & \text{if } n \text{ is odd} \end{cases}$
$C_m \boxtimes C_n$	$\begin{cases} 3, & \text{if } m = n \text{ is odd} \\ 2, & \text{otherwise} \end{cases}$	$\begin{cases} 4, & \text{if } m \text{ and } n \text{ are even} \\ 4-6, & \text{if } m \text{ is even and } n \text{ is odd} \\ 5-7, & \text{if } m \text{ and } n \text{ are odd} \end{cases}$

Certainly, cycles are graphs without simplicial vertices, and hence they are not extreme geodesic graphs. This means that the calculation of the geodetic and the hull numbers of strong product graphs of the form $G \boxtimes C_n$ requires a different approach to the previous one. In Table 3.3, both the hull and the geodetic number of some basic families of the form $G \boxtimes C_n$ is displayed [27].

Theorem 3.9 ([158]). *A nontrivial set $C \subseteq V(G \boxtimes H)$ is convex if and only if the following conditions hold:*

1. *C is two-convex.*
2. *For every $(g,h),(g',h) \in C$, $I_G[g,g'] \times \{h\} \subseteq C$.*
3. *For every $(g,h),(g,h') \in C$, $\{g\} \times I_H[h,h'] \subseteq C$.*
4. *For every $(g,h),(g',h') \in C$, if $d_G(g,g') = d_H(h,h')$, then $I_{G \boxtimes H}[(g,h),(g',h')] \subseteq C$.*
5. *Both $p_G(C)$ and $p_H(C)$ are convex in G and H, respectively.*

It remains an **open problem** to obtain tight lower and upper bounds for $\mathrm{con}(G \boxtimes H)$.

Lemma 3.1 ([27]). *Let $u = (g,h), v = (g',h') \in V(G \boxtimes H)$ such that $d_{G \boxtimes H}(u,v) = d_G(g,g') = l$. If γ is a $(g,h) - (g',h')$ geodesic, then the projection of γ onto G is a $g - g'$ geodesic of length l.*

Proof. If $V(\gamma) = \{(g,h),(g_1,h_1),\ldots,(g_{l-1},h_{l-1}),((g',h'))\}$, then its projection into G is $p_G(V(\gamma)) = \{g,g_1,\ldots,g_{l-1},g'\}$. Since $d_{G \boxtimes H}((g,h),(g',h')) = d_G(g,g')$, $p_G(V(\gamma))$ does not contain repeated vertices, which means that every pair of consecutive vertices are adjacent, i.e., $p_G(V(\gamma))$ is the vertex set of a $g - g'$ geodesic in G.   □

Lemma 3.2 ([27]). *Let $S_1 \times S_2 \subseteq V(G \boxtimes H)$ be a set of vertices of cardinality 6, where $S_1 = \{g_1,g_2\} \subseteq V(G)$ and $S_2 = \{h_1,h_2,h_3\} \subseteq V(H)$. Then*

(1) $(g_2,h_2) \notin I[\{(g_1,h_1),(g_1,h_2),(g_2,h_1)\}]$.
(2) If $h_3 \notin I[h_1,h_2]$, then $(g_2,h_3) \notin I[\{(g_1,h_1),(g_1,h_2)\}]$.
(3) If $h_3 \notin I[h_1,h_2]$, then $(g_1,h_3) \notin I[\{(g_1,h_1),(g_2,h_2)\}]$.
(4) If $h_1 \in I[h_2,h_3]$, then $h_2 \notin I[h_1,h_3]$ and $h_3 \notin I[h_1,h_2]$.

Proof.

(1) Observe that $d((g_1,h_1),(g_1,h_2)) = d(h_1,h_2)$. Hence, according to Lemma 3.1, every $(g_1,h_1) - (g_1,h_2)$ geodesic may not pass through (g_2,h_2), which means that $(g_2,h_2) \notin I[\{(g_1,h_1),(g_1,h_2)\}]$. Similarly, it is proved that $(g_2,h_2) \notin I[\{(g_1,h_1),(g_2,h_1)\}]$. Finally, suppose w.l.o.g that $d((g_1,h_2),(g_2,h_1)) = d(g_1,g_2)$. Hence, again according to Lemma 3.1, every $(g_1,h_2) - (g_2,h_1)$ geodesic may not pass through (g_2,h_2), which means that $(g_2,h_2) \notin I[\{(g_1,h_2),(g_2,h_1)\}]$.

(2) Observe that $d((g_1,h_1),(g_1,h_2)) = d(h_1,h_2)$. Hence, according to Lemma 3.1, the projection onto H of a $(g_1,h_1) - (g_1,h_2)$ geodesic passing through (g_2,h_3) is a $h_1 - h_2$ geodesic passing through h_3, contradicting the hypothesis $h_3 \notin I[h_1,h_2]$.

(3) Suppose that $(g_1,h_3) \in I[\{(g_1,h_1),(g_2,h_2)\}]$.

   If $d((g_1,h_1),(g_2,h_2)) = d(g_1,g_2)$, then, according to Lemma 3.1, every $(g_1,h_1) - (g_2,h_2)$ geodesic may not pass through (g_1,h_3).

   If $d((g_1,h_1),(g_2,h_2)) = d(h_1,h_2)$, then the projection onto H of a $(g_1,h_1) - (g_2,h_2)$ geodesic passing through (g_1,h_3) is an $h_1 - h_2$ geodesic passing through h_3, which contradicts the hypothesis $h_3 \notin I[h_1,h_2]$.

(4) Assume, to the contrary, that, for example, $h_2 \in I[h_1,h_3]$. Then, if $d(h_1,h_2) = x$, $d(h_1,h_3) = y$, and $d(h_2,h_3) = z$, we have that $x + y = z$ and $x + z = y$, i.e., $d(h_1,h_2) = 0$, a contradiction. □

Proposition 3.1 ([27]). *Let G and H be nontrivial graphs. Then, $g(G \boxtimes H) \geq 4$.*

Proof. Let us see that every subset S of $V(G \boxtimes H)$ having at most 3 vertices is not geodetic. Suppose, to the contrary, that S is a geodetic set of cardinality 3. Without loss of generality, we may assume that $|p_G(S)| \leq |p_H(S)|$. We consider different cases.

Case 1: $|p_G(S)| = 1$: In other words, $S = \{(g_1,h_1),(g_1,h_2),(g_1,h_3)\}$ and $|p_H(S)| = 3$. According to Lemma 3.2(4), we may assume w.l.o.g that $h_3 \notin I[h_1,h_2]$, and from Lemma 3.2(1, 2), we conclude that $(g_2,h_3) \notin I[S]$ for any vertex $g_2 \neq g_1$ (see Fig. 3.1a).

Case 2: $|p_G(S)| = |p_H(S)| = 2$: In other words, $S = \{(g_1,h_1),(g_1,h_2),(g_2,h_1)\}$, being $g_1 \neq g_2$ and $h_1 \neq h_2$. From Lemma 3.2(1), we conclude that $(g_2,h_2) \notin I[S]$ (see Fig. 3.1b).

Case 3: $|p_G(S)| = 2$ and $|p_H(S)| = 3$: In other words, $S = \{(g_1,h_1),(g_1,h_2),(g_2,h_3)\}$, being $g_1 \neq g_2$ and h_1,h_2,h_3 three different vertices of H. According to Lemma 3.2(4), we may assume w.l.o.g that $h_1 \notin I[h_2,h_3]$. From Lemma 3.2(1,2), we conclude that $(g_2,h_1) \notin I[S]$ (see Fig. 3.1c).

Case 4: $|p_G(S)| = |p_H(S)| = 3$: In other words, $S = \{(g_1,h_1),(g_2,h_2),(g_3,h_3)\}$, being g_1,g_2,g_3 three different vertices of G and h_1,h_2,h_3 three different vertices of H. According to Lemma 3.2(4), we may assume w.l.o.g that $h_1 \notin I[h_2,h_3]$ and $g_3 \notin I[g_1,g_2]$. From Lemma 3.2(1, 2), we conclude that $(g_3,h_1) \notin I[S]$ (see Fig. 3.1d). □

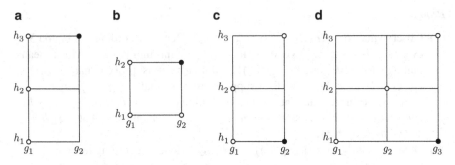

Fig. 3.1 In each figure, the black vertex does not belong to the geodetic closure of white vertices

Proposition 3.2 ([27]). *Let G, H be connected graphs and let $S_1 \subseteq V(G)$, $S_2 \subseteq V(H)$ and $S \subseteq V(G \boxtimes H)$.*

1. *For every integer $r \geq 1$, $I^r[S_1] \times I^r[S_2] \subseteq I^r[S_1 \times S_2]$.*
2. *If S_1 is a geodetic (resp. hull) set of G and S_2 is a geodetic (resp. hull) set of H, then $S_1 \times S_2$ is a geodetic (resp. hull) set of $G \boxtimes H$.*
3. *If S is a geodetic set of $G \boxtimes H$, then either $p_G(S)$ is a geodetic set of G or $p_H(S)$ is a geodetic set of H.*

Proof.

1. We proceed by induction on r. Suppose that $r = 1$ and take a vertex $(g, h) \in I[S_1] \times I[S_2]$. Since $g \in I[S_1]$, then $g \in I[g', g'']$ for some $g', g'' \in V(S_1)$ and thus $d(g', g'') = d(g', g) + d(g, g'')$. Similarly, $d(h', h'') = d(h', h) + d(h, h'')$ for some $h', h'' \in V(S_2)$. We may assume without loss of generality that $d(g', g) \leq d(g, g'')$, $d(h', h) \leq d(h, h'')$, and $d(g', g) \leq d(h', h)$. Then, $d((g', h'), (g, h)) = d(h', h)$ and $d((g, h), (g', h'')) = d(h, h'')$, which means that

$$d((g', h'), (g', h'')) = d(h', h'') = d(h', h) + d(h, h'')$$
$$= d((g', h'), (g, h)) + d((g, h), (g', h'')).$$

   In other words, $(g, h) \in I[(g', h'), (g', h'')] \subseteq I[S_1 \times S_2]$.
   Assume then that $r > 1$. By the inductive hypothesis, $I^{r-1}[S_1] \times I^{r-1}[S_2] \subseteq I^{r-1}[S_1 \times S_2]$. Hence, $I^r[S_1] \times I^r[S_2] = I[I^{r-1}[S_1]] \times I[I^{r-1}[S_2]] \subseteq I[I^{r-1}[S_1] \times I^{r-1}[S_2]] \subseteq I[I^{r-1}[S_1 \times S_2]] = I^r[S_1 \times S_2]$.
2. Let r, s be positive integers such that $I^r[S_1] = V(G)$ and $I^s[S_2] = V(H)$. We may suppose w.l.o.g that $r \leq s$. Then, $V(G \boxtimes H) = V(G) \times V(H) = I^s[S_1] \times I^s[S_2] \subseteq I^s[S_1 \times S_2]$.
3. Assume that neither $S_1 = p_G(S)$ nor $S_2 = p_H(S)$ is geodetic and consider $g \in V(G) \backslash I[S_1]$ and $h \in V(H) \backslash I[S_2]$. As $(g, h) \in I[S] = V(G \boxtimes H)$, then $(g, h) \in I[(g', h'), (g'', h'')]$ for some $(g', h'), (g'', h'') \in S$. Hence, $d((g', h'), (g'', h'')) = d((g', h'), (g, h)) + d((g, h), (g'', h''))$.

On the other hand, as $g \notin I[g',g'']$ and $h \notin I[h',h'']$, we have that $d(g',g'') < d(g',g)+d(g,g'')$ and $d(h',h'') < d(h',h)+d(h,h'')$. Hence,

$$\max\{d(g',g''),d(h',h'')\} < \max\{d(g',g)+d(g,g''),d(h',h)+d(h,h'')\}$$
$$\leq \max\{d(g',g),d(h',h)\}+\max\{d(g,g''),d(h,h'')\}$$
$$= d((g',h'),(g,h))+d((g,h),(g'',h''))$$

which contradicts the previous expression for the distance between (g',h') and (g'',h''). $\qquad\square$

Theorem 3.10. *Let G,H be connected graphs. Then, $\min\{g(G),g(H)\} \leq g(G \boxtimes H) \leq g(G)g(H)$. Furthermore, both bounds are sharp.*

Proof. First, we prove the upper bound. Let S_1 and S_2 be geodetic sets of G and H with minimum cardinality, that is, such that $|S_1| = g(G)$ and $|S_2| = g(H)$. By Proposition 3.2, $S_1 \times S_2$ is a geodetic set of $G \boxtimes H$ with cardinality $|S_1 \times S_2| = |S_1||S_2| = g(S_1)g(S_2)$. Hence, $g(G \boxtimes H) \leq g(G)g(H)$.

To prove the lower bound, take a minimum geodetic set S of $G \boxtimes H$. According to Proposition 3.2, we may suppose, without loss of generality, that $p_G(S)$ is a geodetic set of G. Hence, $\min\{g(G),g(H)\} \leq g(G) \leq |p_G(S)| \leq |S| = g(G \boxtimes H)$.

To show the sharpness of the upper bound, take $G = K_m$ and $H = K_n$. Then, $g(K_m \boxtimes K_n) = g(K_{mn}) = mn = g(K_m)g(K_n)$. Finally, to show the sharpness of the lower bound, take $G = K_{r,s}$, $H = K_n$, with $r,s,n \geq 4$, and check that $g(K_{r,s} \boxtimes K_n) = 4 = g(K_{r,s}) = \min\{g(K_{r,s}),g(K_n)\}$. $\qquad\square$

Theorem 3.11. *For any two nontrivial graphs G and H, $2 \leq h(G \boxtimes H) \leq h(G)h(H)$. Furthermore, both bounds are sharp.*

Proof. First, we prove the upper bound. Let S_1 and S_2 be hull sets of G and H with minimum cardinality, that is, such that $|S_1| = h(G)$ and $|S_2| = h(H)$. By Proposition 3.2, $S_1 \times S_2$ is a hull set of $G \boxtimes H$ with cardinality $|S_1 \times S_2| = |S_1||S_2| = h(S_1)h(S_2)$. Hence, $h(G \boxtimes H) \leq h(G)h(H)$.

To prove the sharpness of this bound, take $G = K_m$ and $H = K_n$ and notice that $h(K_m \boxtimes K_n) = h(K_{mn}) = mn = h(K_m)h(K_n)$.

Finally, the lower bound is a direct consequence of the fact that $h(G) = 1$ if and only if $G = K_1$. As for its sharpness, it is straightforward to check that $\{(0,0),(0,2)\}$ is a hull set of $P_2 \boxtimes C_4$ (see Fig. 3.2). $\qquad\square$

Remark 3.1. In contrast to the geodetic case, the claim $\min\{h(G),h(H)\} \leq h(G \boxtimes H)$ is far from being true in general. A simple counterexample is $C_5 \boxtimes C_7$, as it is straightforward to check that $h(C_5) = h(C_7) = 3$ and $h(C_5 \boxtimes C_7) = 2$.

Fig. 3.2 $P_2 \boxtimes C_4$

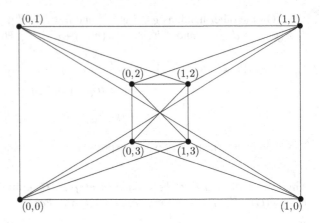

3.3   Lexicographic Product

The *lexicographic product* of graphs G and H is the graph $G \circ H$ on vertex set $V(G) \times V(H)$ in which vertices (g_1, h_2) and (g_2, h_2) are adjacent if and only if either $g_1 g_2 \in E(G)$ or $g_1 = g_2$ and $h_1 h_2 \in E(H)$. This graph operation is also known as the graph *composition* and denoted by $G[H]$. The graph $G \circ H$ is called nontrivial if both factors are graphs on at least two vertices. Next, we show a basic list of properties of this graph operation, whose proofs are direct consequences of the definition.

Theorem 3.12. *Let G, H be graphs.*

1. *The graph $G \circ H$ is connected if and only if G is connected.*
2. *The lexicographic product is associative but not commutative.*
3. *If G is connected, then*

   - $d_{G \circ H}((g,h),(g',h')) = d_G(g,g')$ *if $g \neq g'$.*
   - $d_{G \circ H}((g,h),(g,h')) = 2$ *if $hh' \in E(H)$.*
   - $d_{G \circ H}((g,h),(g,h')) = 1$ *if $hh' \in E(H)$.*

4. $\omega(G \circ H) = \omega(G)\omega(H)$.
5. $\text{Ext}(G \circ K_q) = \text{Ext}(G) \times V(K_q)$.

Lemma 3.3. *Let W be a proper convex set of a connected lexicographic product $G \circ H$. If ρ is an induced $(g,h) - (g',h')$ path of length 2 in W, then $g \neq g'$.*

Proof. Suppose, to the contrary, that $g = g'$. Hence, $hh' \notin V(H)$ and, for every $x \in N_G(g)$, $\{x\} \times V(H) \subseteq W$. Since $d_{G \circ H}((x,h),(x,h')) = 2$, for every $y \in N_G(x)$, $\{y\} \times V(H) \subseteq W$. As G is connected, by induction on the distance from g, we conclude that $W = V(G \circ H)$, a contradiction.   □

Theorem 3.13 ([1]). *Let W be a proper subset of a nontrivial connected lexicographic product $G \circ H$. If W induces a non-complete subgraph of $G \circ H$, then W is convex if and only if the following conditions hold:*

1. $p_G(W)$ *is convex in* G.
2. $\{g\} \times V(H) \subseteq W$, *for any non-simplicial vertex* $g \in p_G(W)$.
3. H *is a complete graph.*

Proof. Suppose first that 1, 2, 3 hold. Since H is complete, any vertices (g, h_1) and (g, h_2) that belong to W are adjacent and hence clearly $I_{G \circ H}[(g, h_1), (g, h_2)] \subseteq W$. Consider next a pair of vertices $(g_1, h_1), (g_2, h_2) \in W$, where $g_1 \neq g_2$. Let P be a $(g_1, h_1) - (g_2, h_2)$ a geodesic in $G \circ H$. Then, $p_G(P)$ is $g_1 - g_2$ geodesic in G. Let (g, h) be an arbitrary inner vertex of P. Since $g \in V(p_G(P))$, this vertex is non-simplicial, which means that $\{g\} \times V(H) \subseteq W$, and in particular, $(g, h) \in W$.

Conversely, take a proper convex set W of $G \circ H$ whose induced subgraph is not a clique. Let $g_0, g_r \in p_G(W)$ and ρ a $g_0 - g_r$ geodesic such that $V(\rho) = \{g_i\}_{i=0}^r$. Clearly, for every $h \in V(H)$, the set $V(\rho)$ induces a $(g_0, h) - (g_r, h)$ geodesic P in $G \circ H$. Since $(g_0, h), (g_r, h) \in W$ and W is convex, $\{(g_i, h)\}_{i=0}^r \subseteq W$, which means that $V(\rho) = \{g_i\}_{i=0}^r \subseteq p_G(W)$. Thus item 1 holds.

Next, to prove that H is a complete graph, suppose, to the contrary, that $h, h' \in V(H)$ and $hh' \notin E(H)$. Take $g, g' \in V(G)$ such that $gg' \in E(G)$. Notice that the set $S = \{(g, h), (g', h), (g, h')\}$ induces a path of length 2, contradicting Lemma 3.3. Hence, item 3 holds.

Finally, let g be a non-simplicial vertex of $p_G(W)$. Take $g', g'' \in N_G(g) \cap p_G(W)$ such that $d_G(g', g'') = 2$. Clearly, for every $h \in V(H)$, $d_{G \circ H}((g', h), (g'', h)) = 2$. Hence, as W is convex and $\{g\} \times V(H)$ induces a clique, $\{g\} \times V(H) \subseteq W$. □

Corollary 3.3. *Let* $G \circ H$ *be a nontrivial connected lexicographic product. Then,* $\text{con}(G \circ H) = \omega(G \circ H)$, *unless* H *is a complete graph.*

Theorem 3.14 ([35]). *Let* G *be a connected graph of order* p. *If* G *has no simplicial vertex, then* $\text{con}(G \circ K_q) = q \cdot \text{con}(G)$; *otherwise,* $\text{con}(G \circ K_q) = pq - 1$.

Proof. Suppose first that $\text{Ext}(G) = \emptyset$. Let W be a proper convex of $\text{con}(G \circ H)$. Certainly, W induces a clique of $G \circ K_q$ if and only if $p_G(W)$ induces a clique of G. Observe also that, if W does not induce a clique, according to Theorem 3.13, $p_G(W)$ is a proper convex set of G. Hence, $\text{con}(G \circ K_q) \leq q \cdot \text{con}(G)$. As for the other inequality, it is a direct consequence of the fact that, for every convex set C of G, the set $C \times V(K_q)$ is a convex set of $\text{con}(G \circ H)$.

As for case $\text{Ext}(G) \neq \emptyset$, the equality $\text{con}(G \circ K_q) = pq - 1$ is directly derived from Proposition 2.2 and Theorem 3.12(5). □

Lemma 3.4. *Let* G *be a non-complete connected graph and* $q \geq 2$. *If* $A \subseteq V(G \circ K_q)$ *does not induce a clique and* $k \geq 1$, *then*

1. $p_G(I_{G \circ K_q}^k[A]) = I_G^k[p_G(A)]$.
2. $I_{G \circ K_q}^k[A] \subseteq I_G^k[p_G(A)] \times V(K_q)$.

Proof.

1. Assume first that $k = 1$ and take $(g, h) \in I_{G \circ K_q}[A]$. If P is a $(g_1, h_1) - (g_2, h_2)$ geodesic such that $(g, h) \in V(P)$ and $(g_1, h_1), (g_2, h_2) \in A$, then $p_G(P)$ induces a $g_1 - g_2$ geodesic in G containing g, which means that $g \in I_G[p_G(A)]$.

Fig. 3.3 Graph $C_5 \circ K_2$. Note that $h(C_5 \circ K_2) = 3$ and $g(C_5 \circ K_2) = 5$

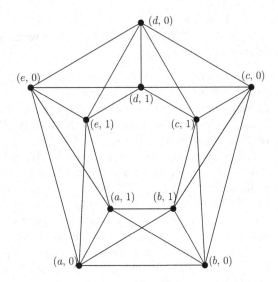

Conversely, given an arbitrary vertex $g \in I_G[p_G(A)]$, if ρ is a $g_1 - g_2$ geodesic ρ such that $g_1, g_2 \in P_G(A)$ and $g \in V(\rho)$, then there exists a $(g_1, h_1) - (g_2, h_2)$ geodesic P in $G \circ K_q$ such that $(g_1, h_1), (g_2, h_2) \in A$ and $(g, h) \in V(P)$, which means that $(g, h) \in I_{G \circ K_q}[A]$, i.e., $g \in p_G(I_{G \circ K_q}[A])$. Proceeding by induction on k, we obtain the desired result $I_G^k[p_G(A)] = I_G[I_G^{k-1}[p_G(A)]] = I_G[p_G(I_{G \circ K_q}^{k-1}[A])] = p_G(I_{G \circ K_q}[I_{G \circ K_q}^{k-1}[A]]) = p_G(I_{G \circ K_q}^k[A])$.

2. $I_{G \circ K_q}^k[A] \subseteq p_G(I_{G \circ K_q}^k[A]) \times p_H(I_{G \circ K_q}^k[A]) \subseteq I_G^k[p_G(A)] \times V(K_q)$. □

It is significant to observe that containment in item 2 in the previous lemma may be strict. For example, the set $A = \{(a, 1), (b, 1), (d, 1)\}$ of $H = C_5 \circ K_2$ satisfies $I_H[A] = V(H) \setminus \{(a, 0), (b, 0), (d, 0)\}$ and $I_{C_5}[p_{C_5}(A)] \times V(K_2) = I_{C_5}[\{a, b, d\}] \times V(K_2) = V(H)$ (see Fig. 3.3).

Theorem 3.15 ([34]). *If G is a non-complete connected graph and $q \geq 2$, then $h(G \circ K_q) = h(G) + (q - 1) \cdot |\mathrm{Ext}(G)|$.*

Proof. Take a minimum hull set A of $G \circ K_q$. Notice that $V(G \circ K_q) = [A] = I_{G \circ K_q}^k[A]$, for some integer $k \geq 1$. Thus, according to Lemma 3.4, $p_G(A)$ is a hull set of G, which means that $h(G \circ K_q) = |A| \geq h(G) + (q - 1) \cdot |\mathrm{Ext}(G)|$, since $\mathrm{Ext}(G) \subseteq p_G(A)$ and $\mathrm{Ext}(G) \times V(K_q) = \mathrm{Ext}(G \circ K_q) \subseteq A$.

Conversely, take a minimum hull set B of G and an arbitrary vertex $u \in V(K_q)$. Consider the set $A = (B \times \{u\}) \cup \mathrm{Ext}(G \circ K_q)$. Observe that $p_G(A) = B$ and $|A| = |B \setminus \mathrm{Ext}(G)| + |\mathrm{Ext}(G \circ K_q)| = h(G) + (q - 1) \cdot |\mathrm{Ext}(G)|$ since $\mathrm{Ext}(G) \subseteq B$. Notice also that $V(G) \times \{u\} \subseteq [A]$. Take a vertex $(g, h) \in V(G \circ K_q)$ such that $g \notin \mathrm{Ext}(G)$ and $h \neq u$. Take a pair of vertices $g_1, g_2 \in N_G(g)$ such that $g_1 g_2 \notin E(G)$. Clearly, $(g_1, u), (g_2, u)$ are two nonadjacent neighbors of (g, u) in $G \circ K_q$, and hence, they are also neighbors of vertex (g, h). In consequence, $(g, h) \in I_{G \circ K_q}[(g_1, u), (g_2, u)] \subseteq [A]$.

So, we have proved that A is a hull set of cardinality $h(G) + (q-1) \cdot |\text{Ext}(G)|$, which means that $h(G \circ K_q) \leq |A| = h(G) + (q-1) \cdot |\text{Ext}(G)|$. $\qquad\square$

Theorem 3.16 ([30]). *If G is a non-complete connected graph and $q \geq 2$, then*

$$\max\{4, q \cdot |\text{Ext}(G)|\} \leq g(G \circ K_q) \leq \min\{q \cdot g(G), (q-1) \cdot |\text{Ext}(G)| + |V(G)|\}.$$

Moreover, both bounds are tight.

Proof. Let $H = G \circ K_q$. First, we prove the lower bound.

Let $W = \{(g_1, h_1), (g_2, h_2), (g_3, h_3)\}$ be a geodetic set of $G \circ K_q$. Clearly $|g_1, g_2, g_3| = 3$, as otherwise, if $g_1 = g_2$, then every vertex of the form (g_3, h') with $h' \neq h_3$ is outside $I_H[W]$. Take a vertex $h' \in V(K_q)$ such that $h' \neq h_3$. Then, $d_H((g_1, h_1), (g_3, h_3)) = d_H((g_1, h_1), (g_3, h'))$ and $d_H((g_2, h_2), (g_3, h_3)) = d_H((g_2, h_2), (g_3, h'))$, which means that $(g_3, h') \in I_H[(g_1, h_1), (g_2, h_2)]$. As $d_H((g_1, h_1), (g_2, h_2)) = d_G(g_1, g_2)$, $d_H((g_1, h_1), (g_3, h')) = d_G(g_1, g_3)$, and $d_H((g_3, h'), (g_2, h_2)) = d_G(g_1, g_3)$, it follows that $(g_3, h_3) \in I_H[(g_1, h_1), (g_2, h_2)]$. Hence, $W = \{(g_1, h_1), (g_2, h_2)\}$ is a geodetic set of $G \circ K_q$. Take a vertex $h' \in V(K_q)$ such that $h' \neq h_1$. Then, $d((g_1, h_1), (g_2, h_2)) = d((g_1, h_1), (g_2, h'))$, which means that $(g_2, h') \notin I_{G \circ K_q}[(g_1, h_1), (g_2, h_2)]$, a contradiction. Hence, we have seen that $4 \leq g(G \circ K_q)$.

As every extreme vertex must belong to every geodetic set, it follows, according to Theorem 3.12(5), that $q \cdot |\text{Ext}(G)|\} \leq g(G \circ K_q)$.

To show the sharpness of this lower bound take G isomorphic either to a non-complete extreme geodesic graph G or to P_3.

Now, we prove the upper bound. Let B be a minimum geodetic set of G. Consider the set $A = B \times V(K_q)$ and notice that $|A| = q \cdot g(G)$. Take a vertex $(g, h) \in V(G \circ K_q))$ such that $g \notin B$ and a pair of vertices $g_1, g_2 \in B$ such that $g \in I_G[g_1, g_2]$. Hence $(g, h) \in I_H[(g_1, h), (g_2, h)]$, which means that $(g, h) \in I_H[A]$, as $(g_1, h), (g_2, h) \in A$. So, $g(G \circ K_q) \leq q \cdot g(G)$.

Take a vertex $u \in V(K_q)$ and consider the set $C = \text{Ext}(G \circ K_q) \cup (V(G)) \setminus \text{Ext}(G)) \times \{u\}$. Take a vertex $(g, h) \in V(G \circ K_q))$ such that $g \notin \text{Ext}(G)$ and a pair of nonadjacent vertices $g_1, g_2 \in N_G(g)$. Hence $(g_1, u), (g_2, u) \in A \cap N_H((g, h))$ and they are not adjacent, which means that $(g, h) \in I_H[A]$. So, $g(G \circ K_q) \leq |A| = (q-1) \cdot |\text{Ext}(G)| + |V(G)|$.

To show the sharpness of this upper bound take G isomorphic either to a non-complete extreme geodesic graph G or to C_5. $\qquad\square$

Theorem 3.17 ([20, 34]). *Let $G \circ H$ be a nontrivial connected lexicographic product. If H is non-complete, then*

1. $h(G \circ H) = 2$.
2. $2 \leq g(G \circ H) \leq 3 \cdot g(G)$.
3. $g(G \circ H) = 2$ if and only if G contains a vertex u such that $N_G[u] = V(G)$, $g(H) = 2$ and $\text{diam}(H) = 2$.

Proof.

1. Let $x_0 \in V(G)$ and $h, h' \in V(H)$ such that $hh' \notin E(H)$. Take the set $S_0 = \{(x_0, h), (x_0, h')\}$ and note that $N_G(x_0) \times V(H) \subseteq [S_0]$, as $d_{G \circ H}((x_0, h), (x_1, h')) = 2$. Take $x_1 \in N_G(x_0)$ and consider the set $S_0 = \{(x_1, h), (x_1, h')\} \subseteq [S_0]$. Observe that $N_G(x_1) \times V(H) \subseteq [S_1] \subseteq [S_0]$. As G is connected, proceeding by induction on the distance from x_0, we conclude that $[S_0] = V(G \circ H)$, which means that $h(G \circ H) = 2$.

2. Let D be a minimum geodetic set of G. Then, $D = \text{Ext}(G) \cup C$ with $\text{Ext}(G) \cap C = \emptyset$. Take a pair of vertices $h, h' \in V(H)$ such that $hh' \notin E(H)$. For each vertex $g \in \text{Ext}(G)$, choose a vertex $x_g \in N_G(g)$. For each vertex $g \in C$, choose two vertices $y_g, z_g \in N_G(g)$ such that $y_g z_g \notin E(G)$. Consider the set $D' = A_1 \cup A_2 \cup A_3$, where $A_1 = \{(g, h)\}_{g \in V(G)}$, $A_2 = \{(x_g, h), (x_g, h'),\}_{g \in V(G)}$, and $A_3 = \{(y_g, h), (z_g, h),\}_{g \in V(G)}$. Notice that D' is a geodetic set of $G \circ H$. Hence, $g(G \circ H) \leq |D'| \leq 3 \cdot |D| = 3 \cdot g(G)$.

3. Let u a vertex of G such that $N_G[u] = V(G)$. Let $\{h_1, h_2\}$ be a minimum geodetic set of H. Clearly, $d_H(h_1, h_2) = 2$, since $\text{diam}(H) = 2$. Consider the set $S = \{(u, h_1), (u, h_2)\}$. Observe that S is a minimum geodetic set of $G \circ H$, as $\text{diam}(G \circ H) = 2$, $d_{G \circ H}((u, h_1), (u, h_2)) = 2$, and every vertex of $G \circ H$ is adjacent both to (u, h_1) and to (u, h_1).

   Conversely, let $\{(g, h_1), (g', h_2)\}$ be a minimum geodetic set of $G \circ H$. Notice that necessarily $g = g'$, since otherwise $d_{G \circ H}((g, h_1), (g', h_2)) = 1$, a contradiction. Hence, $\{h_1, h_2\}$ is a geodetic set of G, as $\{g\} \times V(H)$ induces a graph isomorphic to H. It is also clear that $\text{diam}(H) = 2$, since $d_{G \circ H}((g, h_1), (g, h_2)) = 2$. Finally, g must be adjacent to any other vertex in G, as every vertex of $G \circ H$ is adjacent both to (g, h_1) and to (g, h_1). $\square$

In [20], it was also proved the following result, from which it is derived that the upper bound of the second item in the previous theorem is tight (a *support vertex* of a tree is any neighbor of a leaf of the tree).

Proposition 3.3 ([20]). *If G is a tree on at least five vertices and $g(H) \cdot \text{diam}(H) > 4$, then $g(G \circ H) = 3 \cdot g(G)$ if and only if (i) every support vertex of G has degree 2 and (ii) there are no three distinct support vertices with a common neighbor.*

3.4 Join

A set S of vertices of a connected graph G is a *non-connecting* set in G if it fulfills the following condition: For every pair of vertices $u, v \in V(G) \setminus S$ with $d(u, v) = 2$, $N(u) \cap N(v) \cap S = \emptyset$. Let $\eta(G)$ denote the cardinality of a minimum non-connecting set of G. Observe that $\eta(G) + \omega(G) \leq |V(G)|$.

Lemma 3.5. *Let G be a connected graph. Let S be a proper convex of $G \vee K_q$. Then, either the set $S \cap V(G)$ induces a clique of G or the set $V(G) \setminus S \cap V(G)$ is a non-connecting set of G.*

Proof. Suppose that $S \cap V(G)$ does not induce a clique of G and take $u, v \in S \cap V(G)$ such that $d_G(u,v) = 2$. Notice that every $u - v$ geodesic is a shortest path of length 2 contained in S. Hence, for every vertex $z \in V(G) \setminus S \cap V(G)$, $z \notin N_G(u) \cap N_G(v)$, which means that $V(G) \setminus S \cap V(G)$ is a non-connecting set of G. $\qquad\square$

Theorem 3.18 ([36]). *Let G be a non-complete graph of order p.*

(1) If G is connected, then $\text{con}(G \vee K_q) = p + q - \eta(G)$.
(2) If G is a non-connected graph and $\{G_i\}_{i=1}^h$ is the set of components of G, then $\text{con}(G \vee K_q) = p + q - \min\{\eta(G_i)\}_{i=1}^h$.
(3) If H is a non-complete graph, then $\text{con}(G \vee H) = \omega(G \vee H) = \omega(G) + \omega(H)$.

Proof.

(1) Let S be a proper convex of $G \vee K_q$. Consider the sets $B = S \cap V(G)$ and $C = S \cap V(K_q)$. Suppose that B does not induce a clique of G and take $a, b \in B$ such that $ab \notin E(G)$. Notice that $V(K_q) \subseteq I_{G \vee H}[a,b] \subseteq S$. Hence, according to Lemma 3.5, $S = B \cup V(K_q)$, where $V(G) \setminus B$ is a non-connecting set, and $|S| = |B| + q \leq (p - \eta(G)) + q$. Finally, as (i) any set of the form $S = B \cup V(K_q)$, where B either induces a clique of G or $V(G) \setminus B$ is non-connecting, is a convex set of $G \vee K_q$ and (ii) if B induces a clique of G, then $V(G) \setminus B$ is non-connecting, we conclude that $\text{con}(G \vee K_q) = p + q - \eta(G)$.
(2) This result is similarly proved as in the previous item, just by also noticing that every set S of the form $S = B_i \cup (\cup_{j \neq i} V(G_j)) \cup V(K_q)$, where $V(G_i) \setminus B_i$ is a non-connecting set of G_i, is a convex set of $G \vee K_q$.
(3) This result is a direct consequence of Theorem 2.12, since $\text{diam}(G \vee H) = 2$, $\overline{G \vee H} = \overline{G} + \overline{H}$ is a non-connected graph and both $\overline{G}$ and $\overline{H}$ contain at least a nontrivial component. $\qquad\square$

A *two-geodetic set* of a graph G is a set D of vertices of G such that every vertex on G lies on some geodesic of length 2 joining two vertices of G [147]. A *two-convex set* of a graph G is a set D of vertices of G such that every geodesic of length 2 joining two vertices of G is contained in D [1]. A *two-hull set* of a graph G is a set D such that the smallest two-convex set containing D is $V(G)$. Let $g_2(G)$ and $h_2(G)$ denote the minimum cardinality of a two-geodetic set and a two-hull set, respectively. Notice that $h(G) \leq h_2(G)$ and $g(G) \leq g_2(G)$, since every two-hull set (resp. two-geodetic set) is a hull set (resp. geodetic set). Certainly, the converse of these assertions are far from being true, in general (see Fig. 3.4), unless the graph has diameter 2.

Lemma 3.6. *Let G and H be non-complete graphs. Then,*

(1) $2 \leq g(G \vee H) \leq \min\{g_2(G), g_2(H)\}$.
(2) If $2 \leq g(G \vee H) \leq 3$, then $g(G \vee H) = \min\{g_2(G), g_2(H)\}$.

Proof.

(1) Certainly, $2 \leq g(G \vee H)$, as $G \vee H$ is a nontrivial graph. Let B be a two-geodetic set of G. Take $a, b \in B$ such that $ab \notin E(G)$. Notice that $V(H) \subseteq I_{G \vee H}[a,b]$.

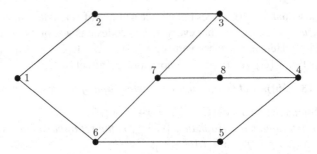

Fig. 3.4 Graph G such that $h(G) = 2$, $h_2(G) = 3$, $g(G) = 3$, $g_2(G) = 4$

Hence, $V(G \vee H) = V(G) \cup V(H) \subseteq I_{G \vee H}[B]$, which means that B is a geodetic set of $G \vee H$.

(2) Suppose that $g(G \vee H) = 3$. Let $S = \{a,b,c\}$ be a minimum geodetic set of $G \vee H$. Notice that either $S \subseteq V(G)$ or $S \subseteq V(H)$. Assume w.l.o.g. that $S \subseteq V(G)$. Next we show that S is a two-geodetic set of G. To this end, take a vertex $u \in V(G) \setminus S$. As S is a geodetic set of $G \vee H$, $u \in I_{G \vee H}[S]$. Assume w.l.o.g. that $u \in I_{G \vee H}[a,b]$. Hence, $d_G(a,b) = 2$ and $a, b \in N_G(u) \cap S$, since $\mathrm{diam}(G \vee H) = 2$, and we are done. Finally, having in mind the previous item, we conclude that $g_2(G) = 3$. The case $g(G \vee H) = 2$ is similarly proved.   □

Theorem 3.19 ([33,36]). *Let G and H be two non-complete graphs. Then,*

(i) $h(G \vee H) = 2$.
(ii) $g(G \vee H) = \min\{g_2(G), g_2(H)\}$.

Proof.

(i) Let $A = \{x,y\} \subseteq V(G)$ such that $xy \notin E(G)$. Clearly, $V(H) \subseteq I_{G \vee H}[A]$. Let $u,v \in V(H)$ such that $uv \notin V(H)$. Notice that $V(G) \subseteq I_{G \vee H}[u,v] \subseteq I_{G \vee H}^2[A]$. Hence, $V(G \vee H) = V(G) \cup V(H) \subseteq I_{G \vee H}^2[A] \subseteq [A]$, i.e., $[A] = V(G \vee H)$ and $h(G \vee H) = 2$.

(ii) According to Lemma 3.6, what remains to be done is proving that $g(G \vee H) \leq 4$. To this end, let $S = \{x,y,u,v\}$ a set of vertices such that $\{x,y\} \subseteq V(G)$, $\{u,v\} \subseteq V(H)$, $xy \notin E(G)$, and $uv \notin V(H)$. Observe that $V(H) \subseteq I_{G \vee H}[x,y]$ and $V(G) \subseteq I_{G \vee H}[u,v]$. Hence, $V(G \vee H) = V(G) \cup V(H) \subseteq I_{G \vee H}[S]$, i.e., $I_{G \vee H}[S] = V(G \vee H)$ and $g(G \vee H) \leq |S| = 4$.   □

Theorem 3.20 ([33,36]). *Let G be a non-complete graph. Then,*

(i) $g(G \vee K_q) = g_2(G)$.
(ii) $h(G \vee K_q) = h_2(G)$.

Proof. First, notice that if a vertex v of a graph H is *universal*, i.e., if $N_H[v] = V(H)$, then it does not belong to any minimum geodetic set (resp. hull set) of H. This means that every minimum geodetic set (resp. minimum hull set) of $G \vee K_q$ is a subset of $V(G)$.

Finally, as $\text{diam}(G \vee K_q) = 2$, it follows that, for every $S \subset V(G)$, the following three statements are equivalent:

- S is a two-geodetic set (resp. two-hull set) of G.
- S is a two-geodetic set (resp. two-hull set) of $G \vee K_q$.
- S is a geodetic set (resp. hull set) of $G \vee K_q$. □

3.5 Corona Product

Let G and H be two graphs and let n be the order of G. The *corona product*, or simply the *corona*, of graphs G and H is the graph $G \odot H$ obtained by taking one copy of G and n copies of H and then joining by an edge the ith vertex of G to every vertex in the ith copy of H [103]. Given a vertex $g \in G$, the copy of H connected to g is denoted by H_g. Complete graphs, stars, wheels, fan graphs, and comb graphs are basic examples of corona product families, all of them except the last one being of the type $K_1 \odot H$. See Fig. 3.5 for some more examples. Next, we show a basic list of properties of this graph operation, whose proofs are direct consequences of the definition.

Theorem 3.21. *Let G, H be graphs.*

1. $|V(G \odot H)| = |V(G)|(|V(H)| + 1)$.
2. *The graph $G \odot H$ is connected if and only if G is connected.*
3. *The graph $G \odot H$ is complete if and only if $G \cong K_1$ and H is complete.*
4. *The corona product is neither associative nor commutative.*
5. *If G is connected, then $\text{diam}(G \odot H) = \text{diam}(G) + 2$.*
6. $\omega(G \odot H) = \omega(H) + 1$.

Theorem 3.22 ([180]). *Let G be a nontrivial connected graph of order n_1. If H is a graph of order n_2, then*

1. $g(G \odot H) = n_1 \cdot g(K_1 \odot H)$ *if H is non-complete, and $n_1 n_2$ otherwise.*
2. $s(G \odot H) = n_1 \cdot n_2$.
3. $g(G \odot H) \leq s(G \odot H)$.
4. $g(K_1 \odot H) \leq s(K_1 \odot H)$.

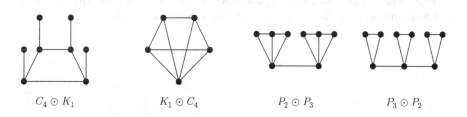

$C_4 \odot K_1$ $\qquad$ $K_1 \odot C_4$ $\qquad$ $P_2 \odot P_3$ $\qquad$ $P_3 \odot P_2$

Fig. 3.5 Some corona graphs

Sketch of proof.

1. If $H \cong K_{n_2}$, then $g(G \odot H)$ is an extreme geodesic graph. Hence, $g(G \odot H) = n_1 n_2$, since $\text{Ext}(G \odot H) = \cup_{g \in G} \text{Ext}(H_g)$.

    Suppose now that H is not a complete graph and take a minimum geodetic set W of $g(G \odot H)$. Check, first, that $W \cap V(G) = \emptyset$ and, second, that for every $x \in V(G)$, $W \cap V(H_g) \neq \emptyset$ is a geodetic set of $\{x\} \odot H_x$. Hence,

$$g(G \odot H) = |W| = \sum_{x \in G} |W \cap V(H_x)| \geq \sum_{x \in G} g(\{x\} \odot H_x) = n_1 \cdot g(K_1 \odot H).$$

    Conversely, let $U_x \subset \{x\} \cup V(H_x)$ be a minimum geodetic set of $\{x\} \odot H_x$. Notice that $x \notin U_x$, i.e., $U_x \subset V(H_x)$. Consider the set $U = \cup_{x \in G} U_x$. Let z be an arbitrary vertex of $G \odot H$. We distinguish cases.

    Case 1: If $z \in V(H_x) \setminus U_x$, then there exist $u, v \in U_x$ s.t. $z \in I_{\{x\} \odot H_x}[u, v]$. So, $z \in I_{G \odot H}[u, v]$.
    Case 2: If $z \in V(G)$ and $n_1 \geq 2$, then for every vertex $u \in U_z$ and some vertex $v \in U_y$, $y \neq z$, $z \in I_{G \odot H}[u, v]$.
    Case 3: If $z \in V(G)$ and $n_1 = 1$, then as H is non-complete, there exist $u, v \in U = U_x$ s.t. $z \in I_{G \odot H}[u, v]$.

    Thus, every vertex z of $G \odot H$ is geodominated by a pair of vertices of U, which means that $g(G \odot H) \leq |U| = n_1 \cdot g(K_1 \odot H)$.
2. Check that $A = \cup_{x \in G} V(H_x)$ is a Steiner set of $G \odot H$. Thus, $s(G \odot H) \leq |A| = n_1 \cdot n_2$.

    Conversely, let B be a minimum Steiner set of $G \odot H$. Check that B must be a subset of A. To prove that $B = A$, suppose, to the contrary, that there exists a vertex $x \in V(G)$ such that $B_x = B \cap V(H_x) \subsetneq V(H_x)$. Take a vertex $u \in V(H_x) \setminus B_x$. Since every vertex of B_x is adjacent to x and x belongs to every Steiner B-tree, we have that the size of the restriction of T to $\{x\} \cup V(H_x)$ is $|B_x|$. Thus, u does not belong to any Steiner B-tree in $G \odot H$, which is a contradiction. Therefore, $s(G \odot H) \geq n_1 \cdot n_2$.
3. Corollary of previous items.
4. Corollary of Theorem 5.5, since $\text{diam}(K_1 \odot H) \leq 2$. $\square$

They remain as *open problems* to obtain tight lower and upper bounds for $h(G \odot H)$ and also for $\text{con}(G \odot H)$.

Further results involving graph operations related to geodesic convexity in graphs can be found in [2, 17, 20, 27, 29–31, 33–36, 74, 126, 130, 158, 178].

Chapter 4
Boundary Sets

Let S be a set of vertices of a connected graph $G = (V,E)$. The *eccentricity* $e_S(v)$ of a vertex $v \in S$ is the maximum distance between v and any other vertex of S, i.e., $e_S(v) = \max\{d(v,u) : u \in S\}$. A vertex $u \in S$ is said to be a contour vertex of S if $e_S(u) \geq e_S(v)$ for every neighbor v of u in S. The set of all contour vertices of S is called the contour set of S and is denoted by $\mathrm{Ct}(S)$.[1]

Theorem 4.1 ([23]). *Let $G = (V,E)$ be a connected graph and $W \subseteq V$ a convex set. Then, W is the convex hull of its contour vertices.*

Proof. Suppose, to the contrary, that $[\mathrm{Ct}(S)] \subsetneq S$. Let $u \in S \setminus [\mathrm{Ct}(S)]$ be such that $e_S(u) \geq e_S(w)$ for all $w \in S \setminus [\mathrm{Ct}(S)]$. Since $u \notin \mathrm{Ct}(S)$, there exists a neighbor v of u in S such that $e_S(v) > e_S(u)$ and, by our choice of u, the vertex v belongs to $[\mathrm{Ct}(S)]$.

Let $v_e \in S$ be an eccentric vertex for v in S, i.e., $d(v,v_e) = e_S(v)$. Note that in this case $e_S(v_e) \geq e_S(v) > e_S(u)$ and $v_e \in [\mathrm{Ct}(S)]$. Therefore, $d(u,v_e) \leq e_S(u) < e_S(v) = d(v,v_e)$ and so $d(u,v_e) + 1 \leq d(v,v_e)$.

Let P be a shortest $v_e - u$ path in S. Then P followed by the edge uv is a $v_e - v$ path whose length is $d(u,v_e) + 1 \leq d(v,v_e)$. So it is a shortest path between v_e and v that contains u. This contradicts the fact that $u \notin [\mathrm{Ct}(S)]$. $\qquad\square$

Corollary 4.1. *The contour $\mathrm{Ct}(G)$ of every connected graph G is a hull set.*

As was pointed out in the [23], the contour of a graph needs not be geodetic. For example, in Fig. 4.1, we illustrate two graphs whose contour set is $\{u,v,w\}$ and $I[\{u,v,w\}] = V \setminus \{z\}$. As for the eccentricity, it is rather easy to design a graph G such that $[\mathrm{Ecc}(G)] \subsetneq V(G)$. For example the tree T such that $V(T) = \{u_i\}_{i=0}^5$ and $E(T) = \{u_i u_{i+1}\}_{i=0}^3 \cup \{u_2 u_5\}$ satisfies $\mathrm{Ecc}(T) = \{u_0, u_4\}$ and $[\mathrm{Ecc}(T)] = V(T) - u_5$.

Lemma 4.1 ([24]). *Let $G = (V,E)$ be a connected graph and $x \in V \setminus \mathrm{Ct}(G)$. Then, there exists a geodesic $\rho(x) = x_0 x_1 x_2 \cdots x_r$ such that $e(x_i) = e(x_{i-1}) + 1$, $i = 1, \ldots, r$ and $x_r \in \mathrm{Ct}(G)$.*

---

[1]It also denotes the subgraph induced by the contour vertices of S.

I.M. Pelayo, *Geodesic Convexity in Graphs*, SpringerBriefs in Mathematics, DOI 10.1007/978-1-4614-8699-2_4, © Ignacio M. Pelayo 2013

a   **b**

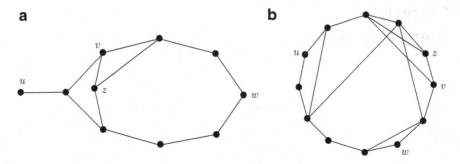

Fig. 4.1 Two graphs of diameter 5 whose contour is not geodetic

Proof. Since the eccentricities of two adjacent vertices differ by at most one unit, if x is not a contour vertex, then there exists a vertex $y \in V$, adjacent to x, such that its eccentricity satisfies $e(y) = e(x) + 1$. This fact implies the existence of a path $\rho(x) = x_0 x_1 x_2 \cdots x_r$, such that $x = x_0$, $x_i \notin Ct(G)$ for $i \in \{0, \ldots, r-1\}$, $x_r \in Ct(G)$, and $e(x_i) = e(x_{i-1}) + 1 = l + i$ for $i \in \{1, \ldots, r\}$, where $l = e(x)$. Moreover, $\rho(x)$ is a shortest $x - x_r$ path, since, otherwise, the eccentricity of x_r would be less than $l + r$. □

Proposition 4.1 ([24, 118]). *Let $G = (V, E)$ be a connected graph.*

1. *If $Ct(G) = Per(G)$, then $I[Ct(G)] = V$.*
2. *If $|Ct(G)| = 2$, then $I[Ct(G)] = V$.*
3. *$I[Ct(G) \cup Ecc(Ct(G))] = V$.*
4. *If $Ecc(Ct(G)) \subseteq Ct(G)$, then $I[Ct(G)] = V$.*
5. *For every $u \in V$, $I[u \cup \partial(u)] = V$.*

Proof.

1. Let x be a vertex of $V(G) \smallsetminus Ct(G)$. According to Lemma 4.1, there exists a vertex $x_r \in Ct(G)$ and a $x - x_r$ geodesic $\rho(x)$ of length r such that $e(x_r) = l + r$, where $l = e(x)$. But $x_r \in Ct(G) = Per(G)$ implies that $e(x_r) = D$ and $D = l + r$. Thus, there exists a vertex $z \in Per(G)$ such that $D = d(z, x_r) \le d(z, x) + d(x, x_r) \le e(x) + r = l + r = D$, that is, $d(z, x_r) = d(z, x) + d(x, x_r)$. Hence, x is on a shortest path between the vertices $z, x_r \in Per(G) = Ct(G)$.
2. If $|Ct(G)| = 2$, then $Ct(G) = Per(G)$. Therefore, according to the previous item, the contour $Ct(G)$ must be geodetic.
3. Let x be a vertex of $V(G) \smallsetminus \Omega(G)$. Since $x \notin Ct(G)$, according to Lemma 4.1, there exists a vertex $x_r \in Ct(G)$ and a $x - x_r$ geodesic $\rho(x)$ of length r such that $e(x_r) = e(x) + r$. Let y_r be an eccentric vertex of x_r, i.e., such that $d(y_r, x_r) = e(x_r)$. Then,

$$e(x) + r = e(x_r) = d(y_r, x_r) \le \underbrace{d(y_r, x)}_{\le e(x)} + \underbrace{d(x, x_r)}_{=r} \le e(x) + r$$

and hence we conclude that the inequalities in the formula above are all of them equalities, which means that the vertex x lies in a shortest path joining $x_r \in \mathrm{Ct}(G) \subset \Omega(G)$ and $y_r \in \mathrm{Ecc}_G(\mathrm{Ct}(G)) \subset \Omega(G)$.

4. Immediate corollary of previous item.
5. Pick any vertex $x \notin u \cup \partial(u)$. Any $u - x$ geodesic extends to a maximal $u - y$ geodesic. This means that $y \in \partial(u)$ and, consequently, x are in a geodesic with both endpoints in $u \cup \partial(u)$. $\qquad\Box$

Corollary 4.2 ([24]). *The boundary $\partial(G)$ of every connected graph $G = (V,E)$ is geodetic.*

Theorem 4.2 ([24]). *Let $G = (V,E)$ be a connected graph such that $\mathrm{Ct}(G) \smallsetminus \mathrm{Per}(G) = \{y_1,\ldots,y_k\}$ and $e(y_i) = e(y_j)$, for each $i,j = 1,\ldots,k$. Then, $I^2[\mathrm{Ct}(G)] = V$.*

Proof. Suppose that $\mathrm{Per}(G) \subsetneq \mathrm{Ct}(G)$ as otherwise by Proposition 4.1 $I[\mathrm{Ct}(G)] = V$. Therefore, if $\mathrm{Per}(G) = \{x_1,\ldots,x_h\}$ and $\mathrm{Ct}(G) = \{x_1,\ldots,x_h,y_1,\ldots,y_k\}$, then $e(x_i) = D$, $i = 1,\ldots,h$, $e(y_j) = d$, $j = i,\ldots,k$, and $d < D$, where D is the diameter of G.

Notice that $\mathrm{Ecc}(\mathrm{Ct}(G)) = \mathrm{Per}(G) \cup \mathrm{Ecc}(\{y_1\}) \cup \cdots \cup \mathrm{Ecc}(\{y_k\})$, as $\mathrm{Ecc}(\mathrm{Per}(G)) = \mathrm{Per}(G)$. Take $v \in \mathrm{Ct}(G) \smallsetminus \mathrm{Per}(G)$. Let w be an eccentric vertex of v, that is, $d(w,v) = e(v)$. Next, we prove that $w \in I[\mathrm{Ct}(G)]$.

Assume that $w \notin \mathrm{Ct}(G)$ since otherwise we are done. Then, by Lemma 4.1, there exists a geodesic $w = w_0 w_1 \cdots w_r$ such that $e(w_i) = e(w_{i-1}) + 1$, $i = 1,\ldots,r$, and $w_r \in \mathrm{Ct}(G)$. Hence, $w_r \in \mathrm{Per}(G)$, since $e(v) \leq e(w) < e(w_r)$. Let z be an eccentric vertex of w_r. Note that $z \in \mathrm{Per}(G)$ and

$$D = d(z,w_r) \leq \underbrace{d(z,w)}_{\leq e(w)} + \underbrace{d(w,w_r)}_{r = e(w_r) - e(w) = D - e(w)} \leq D$$

Thus, w lies in a geodesic joining z and w_r and $\{z,w_r\} \subseteq \mathrm{Per}(G) \subseteq \mathrm{Ct}(G)$. In other words, $w \in I[\mathrm{Ct}(G)]$. We conclude that $\mathrm{Ecc}_G(\mathrm{Ct}(G)) \subset I[\mathrm{Ct}(G)]$ and by Proposition 4.1, $I^2[\mathrm{Ct}(G)] = V$. $\qquad\Box$

As particular cases of the above theorem the following corollary is immediately derived.

Corollary 4.3. *Let $G = (V,E)$ be a connected graph such that either $|\mathrm{Ct}(G)| = |\mathrm{Per}(G)| + 1$ or $|\mathrm{Ct}(G)| = 3$. Then, $I^2[\mathrm{Ct}(G)] = V$.*

Having in mind Theorem 4.2, we examine the geodeticity of the set $I^k[\mathrm{Ct}(G)]$ for some $k \geq 2$. To begin with, we need to introduce a new definition. An integer sequence $(d_1,d_2,\ldots,d_s)$ satisfying $d_1 > d_2 > d > \cdots > d_s$ is called the *eccentricity sequence* of a vertex subset W of a connected graph $G = (V,E)$ if $\{k \in \mathbb{N} : k = e(v)$, for some $v \in W\} = \{d_1,d_2,\ldots,d_s\}$. Moreover, the integer s is called the *size* of the sequence.

Proposition 4.2. *Let $(d_1,d_2,\ldots,d_s)$ be the eccentricity sequence of the contour of a connected graph $G = (V,E)$. Let $x \in \mathrm{Ct}(G)$ and $i \in \{1,\ldots,s\}$ such that $e(x) = d_i$. Then, $\mathrm{Ecc}(\{x\}) \subseteq I^{i-1}[\mathrm{Ct}(G)]$.*

Proof. We proceed by induction on i. First, if $i = 1$, then $e(x) = d_1 = D$, which means that $x \in \text{Per}(G)$. Hence, $\text{Ecc}(\{x\}) \subseteq \text{Ct}(G) = I^0[\text{Ct}(G)]$ since $\text{Ecc}(\text{Per}(G)) = \text{Per}(G)$.

Take $i \in \{2, \ldots, s\}$ and assume (as inductive hypothesis) that, for every vertex $z \in \text{Ct}(G)$ such that $e(z) = d_j$ with $j \in \{1, \ldots, i-1\}$, $\text{Ecc}(\{z\}) \subseteq I^{j-1}[\text{Ct}(G)]$. Let $x \in \text{Ct}(G)$ such that $e(x) = d_i$. Take $y \in \text{Ecc}(\{x\})$, i.e., such that $d(x,y) = d_i$. Suppose that $y \notin \text{Ct}(G)$ as otherwise we are done. According to Lemma 4.1, there exists a geodesic $y = y_0 y_1 \cdots y_r$ such that $e(y_j) = e(y_{j-1}) + 1$, for $j = 1, \ldots, r$ and $y_r \in \text{Ct}(G)$. If $z \in \text{Ecc}(\{y_r\})$, then $y \in I[z, y_r]$ since

$$e(y) + r = e(y_r) = d(z, y_r) \leq \underbrace{d(z,y)}_{\leq e(y)} + \underbrace{d(y,y_r)}_{=r} \leq e(y) + r.$$

Moreover, it is clear that $d_i = e(x) \leq e(y) < e(y_r)$. Hence, we obtain that $e(y_r) = d_j$ for some $j < i$, which, according to the inductive hypothesis, means that $z \in I^{j-1}[\text{Ct}(G)] \subseteq I^{i-2}[\text{Ct}(G)]$. This fact, along with the statements $y \in I[z, y_r]$ and $y_r \in \text{Ct}(G)$ allows us to conclude, as desired, that $y \in I^{i-1}[\text{Ct}(G)]$. $\qquad \square$

As a consequence of this result, Proposition 4.1 and the known fact that in every graph G of diameter D there are at most $\lfloor D/2 \rfloor + 1$ different eccentricities, the following results are immediately derived.

Corollary 4.4 ([24]). *Let $G = (V, E)$ be a connected graph whose contour has an eccentricity sequence of size s. Then,*

1. $\text{Ecc}(\text{Ct}(G)) \subseteq I^{s-1}[\text{Ct}(G)]$.
2. $\Omega(G) \subseteq I^{s-1}[\text{Ct}(G)]$.
3. $I^s[\text{Ct}(G)] = V$.
4. $I^k[\text{Ct}(G)] = V$, *where* $k = \text{gin}(\text{Ct}(G)) \leq \min\{|\text{Ct}(G)| - 1, \frac{D}{2} + 1, s\}$.

Conjecture 4.1. For every connected graph G, $\text{gin}(\text{Ct}(G)) \leq 2$.

Conjecture 4.2. For every connected graph G, $I^2[\text{Ct}(G)] = V(G)$.

A graph G is called *chordal* if every cycle of order at least 4 has a chord, i.e., if it contains no induced cycles of order greater than 3.

Theorem 4.3 ([100]). *The extreme set of every chordal graph $G = (V, E)$ is a monophonic set, i.e., $J[\text{Ext}(G)] = V$.*

Proof. Let $G = (V, E)$ be a chordal graph of order n. We proceed by induction on n. Clearly, for $n \leq 2$ the result trivially holds. Suppose that the statement is true for every chordal graph on fewer than n vertices. Let v be a non-simplicial vertex of G. Take $u_1, u_2 \in N(v)$ such that $u_1 u_2 \notin E$. Let C be a minimal set of vertices of $V \setminus \{u_1, u_2\}$ which meets all $u_1 - u_2$ paths. Clearly, C is a cutset and $v \in C$.

For $i \in \{1, 2\}$, let W_i be the vertex set of the component of $G - C$ which contains u_i and let $G_i = G[W_i \cup C]$. Notice that C is a minimal cutset of $G[W_1 \cup W_2 \cup C]$, which means that $G[C]$ is a complete graph, since, as was proved in [78], in every chordal graph, every minimal cutset of every induced subgraph induces a clique. By the inductive hypothesis, either u_i is simplicial in G_i or u_i lies on a chordless path between simplicial vertices of G_i. In either case, G_i has a simplicial vertex, say z_i, in W_i, for $i = 1, 2$, as $G[C]$ is a clique. Observe that z_i is also a simplicial vertex in G. Since C is a minimal cutset in $G[W_1 \cup W_2 \cup C]$, there is chordless $z_1 - v$ path, say P_1, in $G[W_1 \cup \{v\}]$ and a chordless $z_2 - v$ path, say P_2, in $G[W_2 \cup \{v\}]$. Since C is a cutset, the path $P_1 \cdot P_2$, obtained by concatenating P_1 and P_2, is a chordless path in G joining simplicial vertices and containing v. $\square$

Proposition 4.3. *A vertex x is an extreme vertex with respect to the monophonic convexity in a graph G if and only if it is simplicial.*

Proof. Let x be an extreme vertex with respect to the monophonic convexity, i.e., such that $V - x$ is m-convex. Then, $V - x$ is also g-convex, which means that x is an extreme vertex with respect to the geodesic convexity. Hence, according to Proposition 2.2, x is a simplicial vertex of G.

Conversely, assume that x is a simplicial vertex of G and take a pair of vertices $u, v \in V(G) - x$. Let ρ be an arbitrary $u - v$ path containing vertex x and take the neighbors $\{a, b\}$ of x belonging to $V(\rho)$. Hence, ρ is not an induced path since $ab \in E(G)$. $\square$

Corollary 4.5 ([100]). *The monophonic convexity in a connected graph G is a convex geometry if and only if G is chordal.*

A *Ptolemaic graph* is a chordal graph which contains no 3-fan as induced subgraph. Howorka proved in [121] that a graph is Ptolemaic if and only if it is both chordal and distance-hereditary.

Corollary 4.6 ([100]). *The extreme set of every Ptolemaic graph $G = (V, E)$ is a geodetic set, i.e., $I[\mathrm{Ext}(G)] = V$. In other words, the geodesic convexity in a connected graph G is a convex geometry if and only if G is Ptolemaic.*

Theorem 4.4 ([118]). *The contour $\mathrm{Ct}(G)$ of every connected graph is a monophonic set.*

Proof. Consider a vertex x of G. Suppose that x is not a contour vertex, i.e., x is a vertex of $V(G) \smallsetminus \mathrm{Ct}(G)$. Since the eccentricities of two adjacent vertices differ by at most one unit, if x is not a contour vertex, then there exists a vertex $y \in V(G)$, adjacent to x, such that its eccentricity satisfies $e(y) = e(x) + 1$. This fact implies the existence of a shortest $x_0 - x_r$ path $\rho(x) = (x_0 x_1 x_2 \cdots x_r)$ such that $x = x_0$, $x_i \notin \mathrm{Ct}(G)$ for $i \in \{0, \ldots, r-1\}$, $x_r \in \mathrm{Ct}(G)$, and $e(x_i) = e(x_{i-1}) + 1 = l + i$ for $i \in \{1, \ldots, r\}$, where $l = e(x)$.

Let us now consider those vertices at distance l from x. Suppose that all of them are at a distance less than $l + r$ from x_r. The vertices at a distance less than l from x are at a distance less than $l + r$ from x_r. Hence, the eccentricity of x_r would be less than $l + r$. This implies the existence of a vertex z at a distance exactly l from x and $l + r$ from x_r, and x lies on a shortest path $\Psi = (z \cdots x x_1 \cdots x_r)$ between z and x_r (Fig. 4.2).

Fig. 4.2 Ψ is a shortest
$z - x_r$ path

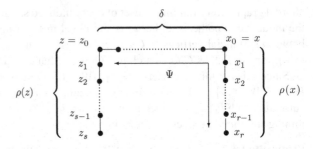

Fig. 4.3 If $d(z_0, x_r) = 2$, then
$l = 1$ and $r = 1$

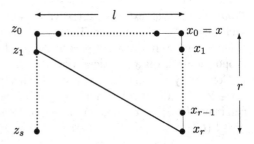

Suppose that z is not a contour vertex, since otherwise we are done. Let us construct a path $\rho(z) = (z_0 z_1 \cdots z_s)$ such that $z = z_0$, $z_i \notin \mathrm{Ct}(G)$ for $i \in \{0, \ldots, s-1\}$, $z_s \in \mathrm{Ct}(G)$, and $e(z_i) = e(z_{i-1}) + 1 = e(z) + i$ for $i \in \{1, \ldots, s\}$ (Fig. 4.2).

Let δ be the $z - x$ subpath of Ψ. Notice that the vertex z satisfies $e(z) \geq l + r$, the vertices of $V(\rho(z)) \smallsetminus \{z\}$ have eccentricity at least $l + r + 1$, and the vertices of $V(\rho(x))$ have eccentricity at most $l + r$. Therefore, the sets $V(\rho(z))$ and $V(\rho(x))$ are disjoint. Moreover, taking into account the eccentricities of all of these vertices, if there is an edge joining a vertex of $V(\rho(z)) \smallsetminus \{z\}$ with a vertex of $V(\rho(x))$, it must be $z_1 x_r$. In this case, $d(z, x_r) = 2 = l + r$, implying that $l = r = 1$ (Fig. 4.3). Hence, the eccentricity of x is 1, the diameter of the graph is 2, and z is a contour vertex, which is a contradiction.

Notice that the sets of vertices $V(\rho(z)) \smallsetminus \{z\}$ and $V(\delta)$ are not necessarily disjoint. Consider a $z_s - x$ path P contained in the walk $\rho(z) \cup \delta = (z_s \cdots z_1 z \cdots x)$. If P has a chord $e = ab$, we can replace the $a - b$ subpath of P with e obtaining a $z_s - x$ path P'. Since $V(P') \subsetneq V(P)$, the path P' has strictly less chords than P. We proceed in an analogous way with P', until we obtain a chordless $z_s - x$ path P^*. Recall that $\Psi = (z \cdots x \cdots x_r)$ is a shortest path, which means that there are no edges joining vertices of $V(\delta) \cup V(\rho(x))$. Therefore, $P^* \cup \rho(x)$ is a monophonic $z_s - x_r$ path through x with $z_s, x_r \in \mathrm{Ct}(G)$. $\qquad\square$

As a consequence of Theorem 4.4, we obtain the following corollary, which was directly proved by Cáceres et al. [23].

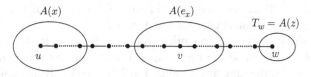

Fig. 4.4 $A(z) = T_w$ is an eccentric vertex of $A(x)$

Corollary 4.7. *The contour of every distance-hereditary graph is a geodetic set.*

Lemma 4.2 ([25]). *Let G be a connected graph and let $u_0 \in V(G)$. Suppose that $(u_0, u_1, \ldots, u_n)$ is a $u_0 - u_n$ path ρ in G such that $e(u_{i+1}) = e(u_i) + 1$, for each $i \in \{0, 1, \ldots, n-1\}$. Then, for each eccentric vertex x of u_n, there exists a $x - u_n$ geodesic that contains ρ.*

Proof. Let x be an eccentric vertex of u_n and suppose that $d(x, u_n) = e(u_n) = k$. Then, by hypothesis, $e(u_0) = k - n$. Let us prove that x is an eccentric vertex of u_0. Suppose, to the contrary, that $d(x, u_0) < k - n$ and consider a $x - u_0$ geodesic between them. Hence, $d(x, u_n) \leq d(x, u_0) + d(u_0, u_n) < k - n + n = k$, which is a contradiction with $d(x, u_n) = k$. Thus, $d(x, u_0) = k - n$ and the path $(x, \ldots, u_0, u_1, \ldots, u_n)$ is the desired geodesic between x and u_n. □

It is well known (see Chap. 2 in [141]) that a connected graph G is chordal if and only if it is isomorphic to the intersection graph of a family $\mathscr{F}$ of subtrees of a tree T, i.e., if there exists a bijection $A : V(G) \mapsto \mathscr{F}$, such that $xy \in E(G)$ if and only if $A(x) \cap A(y) \neq \emptyset$. The pair $(T, \mathscr{F})$ is called a *tree intersection representation* of G and T is the *host tree*. If $x \in V(G)$, we say that the subtree $A(x) \in \mathscr{F}$ *represents* the vertex x. For any vertex $u \in V(T)$, T_u denotes the trivial subtree of T formed only by vertex u. An *l-tree intersection representation* $(T, \mathscr{F}_l)$ is a tree intersection representation satisfying the following property: every leaf u of T belongs to a trivial subtree of $\mathscr{F}_l$. In [25], it was proved that every chordal graph G admits an *l*-tree intersection representation $(T, \mathscr{F}_l)$.

Lemma 4.3 ([25]). *Let $(T, \mathscr{F}_l)$ be an l-tree intersection representation of a chordal graph G. Then,*

1. *For every leaf u of T, the vertex of G represented by T_u is an extreme vertex of G.*
2. *All vertices in G have at least an eccentric vertex which is represented by a trivial subtree T_w, where w is a leaf of T.*

Proof.

1. Let $x \in V(G)$ such that $A(x) = T_u$. All vertices adjacent to x are represented by subtrees meeting T_u, i.e., each of them contains the leaf u, which means that the neighborhood of x is a clique, or equivalently, that $V \smallsetminus \{x\}$ is convex, as desired.
2. Given $x \in V(G)$, let e_x be an eccentric vertex of x, i.e., $d(x, e_x) = e(x) = k$. Let $A(x)$ and $A(e_x)$ be the subtrees of T representing x and e_x, respectively. Assume that $u \in V(A(x))$, $v \in V(A(e_x))$ and consider the path on T between u and v. If v is not a leaf, we can extend the path from v to a leaf w (Fig. 4.4). If v is a leaf,

consider $w = v$. Thus, we obtain a path P of T, from u to a leaf w. Let $z \in V(G)$ be the vertex represented by the subtree $A(z) = T_w$.

Let us show that z is an eccentric vertex of x. Suppose, to the contrary, that $d(x,z) = h < k$. Rename $x = x_0$, $z = x_h$ and take a shortest path $(x_0, x_1, \ldots, x_h)$ in G. Notice that the subgraph L induced in T by vertices lying on the subtrees representing $x_0, x_1, \ldots, x_h$ is connected. Since P is the unique path of T joining $u \in V(A(x_0))$ and $z \in V(A(x_h))$, we have $V(P) \subseteq V(L)$. This means that $v \in V(L)$, so it lies on a subtree representing a vertex x_j, for some $j \in \{0, 1, \ldots, h\}$. Thus, we have $v \in V(A(x_j)) \cap V(A(e_x))$, which implies that x_j and e_x are adjacent.

If $j < h$, there exists a path $(x_0, x_1, \ldots, x_j, e_x)$ of length $j + 1 \leq h < k = e(x)$, which contradicts that e_x is an eccentric vertex of x. Hence, we have that $j = h$, that is, v lies on the subtree $T_w = \{w\}$ representing $x_h = z$, so that $v = w$. Furthermore, since x_{h-1} and $x_h = z$ are adjacent vertices in G, v lies also on the subtree representing x_{h-1}, which meets T_w, and we can build a path $(x_0, x_1, \ldots, x_{h-1}, e_x)$ of length $h < k$, which is again a contradiction.

Therefore, $d(x,z) = k = e(x)$, and z is an eccentric vertex of x, which is represented by a leaf of T.   □

Theorem 4.5 ([25]). *The contour of every chordal graph is a geodetic set.*

Proof. Let G be a chordal graph and let $x_0 \in V(G)$. If $x_0 \notin \mathrm{Ct}(G)$, then it has a neighbor x_1 with greater eccentricity. We repeat this argument to obtain a path $(x_0, x_1, \ldots, x_k)$ where $e(x_{i+1}) = e(x_i) + 1$, $\forall i \in \{0, 1, \ldots, k-1\}$, and x_k is a contour vertex. According to Lemma 4.3(2), there exists an eccentric vertex of x_k, say z, which is represented by a leaf, and by Lemma 4.3(1), such a vertex is a contour vertex. Finally, using Lemma 4.2, we conclude that there exists a shortest path between x_k and z containing x_0, as desired.   □

Certainly, every convex set of a chordal graph induces a chordal subgraph. As a direct consequence, we obtain the following result.

Corollary 4.8. *If G is a chordal graph, then every convex set of G is the geodetic closure of its contour.*

Proposition 4.4 ([96]). *Suppose G is a connected graph with the property that every locally peripheral vertex of G has an eccentric vertex which is also locally peripheral. Then the contour $\mathrm{Ct}(G)$ of G is a geodetic set.*

Proof. Let $v = v_0$ be any vertex of G. If v_0 is not a locally peripheral vertex, then v_0 is adjacent to a vertex v_1 such that $e(v_1) > e(v_0)$. Moreover, if $v_1^{(e)}$ is an eccentric vertex for v_1, then $v_1^{(e)}$ is an eccentric vertex for v_0 and there is a $v_1 - v_1^{(e)}$ geodesic containing v_0. If v_1 is not a locally peripheral vertex, then v_1 is adjacent to a vertex v_2 such that $e(v_2) > e(v_1)$. Moreover, if $v_2^{(e)}$ is an eccentric vertex for v_2, then $v_2^{(e)}$ is an eccentric vertex for v_1 and there is a $v_2 - v_2^{(e)}$ geodesic containing v_1, v_0. Continuing in this manner we construct a sequence $v_0, v_1, \ldots$ of vertices such that $e(v_0) < e(v_1) < \ldots$. This process must terminate with some vertex v_k that is

Fig. 4.5 Some simple graphs

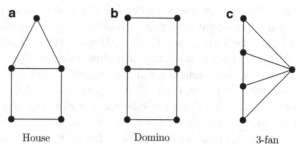

House   Domino   3-fan

necessarily a locally peripheral vertex of G. Moreover, if $v_k^{(e)}$ is an eccentric vertex for v_k which is a locally peripheral vertex, then there is a $v_k - v_k^{(e)}$ geodesic that contains the path $v_k, v_{k-1}, \dots, v_1, v_0$. $\qquad\square$

For an integer $k \geq 2$, a connected graph G is *k-Steiner distance-hereditary* (or simply *k-SDH*), if for every connected induced subgraph H of G and every set S of k vertices of H, $d_H(S) = d_G(S)$, where $d_H(S)$ (resp. $d_G(S)$) denotes the size of a minimum connected subgraph in H (resp. in G) containing S. Clearly, 2-SDH graphs and distance-hereditary graphs constitute the same family. In [105], it was proved that if a graph is k-SDH, then it is t-SDH for all $t \geq k$.

An *HHD-free graph* is a graph in which every cycle of length at least five has at least two chords, i.e., such that it does not contain the house, the domino, and holes (cycles of length at least five) as induced subgraphs (see Fig. 4.5).

Proposition 4.5 ([96]). *If G is either a 3-SDH graph or an HHD-free graph, then every vertex of G has an eccentric vertex which is also locally peripheral.*

Sketch of proof. Consider the following properties:

(A) Every cycle of length at least 6 does not contain adjacent vertices neither of which is incident with a chord.
(B) Every vertex has an eccentric vertex which is locally peripheral.

Prove that (A) implies (B). Finally, prove that if G is either a 3-SDH graph or an HHD-free graph, then every cycle of length at least 6 does not contain adjacent vertices neither of which is incident with a chord. $\qquad\square$

Corollary 4.9 ([96]). *If G is either a 3-SDH graph or an HHD-free graph, then its contour $\mathrm{Ct}(G)$ is a geodetic set.*

A *weakly chordal graph* is a graph such that it does not contain neither holes nor antiholes (complements of holes) as induced subgraphs [14].

Conjecture 4.3. The contour of every weakly chordal graph is a geodetic set.

Proposition 4.6 ([5]). *The contour $\mathrm{Ct}(G)$ of every graph G of diameter at most 4 is a geodetic set.*

Proof. Notice that a vertex y belongs to the geodetic closure of the contour $\mathrm{Ct}(G)$ if and only if there are a pair of vertices $x, z \in \mathrm{Ct}(G)$ such that $d(x,z) = d(x,y) + d(y,z)$. Take a vertex $v \notin I[\mathrm{Ct}(G)]$. Hence, $e(v) < \mathrm{diam}(G) \le 4$. If $\mathrm{diam}(G) = 2$, take a pair of antipodal vertices u, w, thus satisfying $d(u,w) = \mathrm{diam}(G)$. In this case, $e(v) = 1$, which means that $d(u,v) + d(v,w) = 1 + 1 = 2 = d(u,w)$, a contradiction. If $\mathrm{diam}(G) = 3$, then $e(v) = 2$. Take a vertex $w \in N(v)$ such that $e(w) = 3$ and a vertex u such that $d(u,w) = 3$. Hence, $3 = d(u,w) \le d(u,v) + d(v,w) \le 2 + 1 = 3$, i.e., $d(u,w) = d(u,v) + d(v,w)$, a contradiction.

Finally, suppose that $\mathrm{diam}(G) = 4$. Hence, $e(v) \in \{2,3\}$. If $e(v) = 2$, take a pair of antipodal vertices u, w, thus satisfying $d(u,w) = 4$. Hence, $4 = d(u,w) \le d(u,v) + d(v,w) \le 2 + 2 = 4$, i.e., $d(u,w) = d(u,v) + d(v,w)$, a contradiction. If $e(v) = 3$, take a vertex $w \in N(v)$ such that $e(w) = 4$ and a vertex u such that $d(u,w) = 4$. Hence, $4 = d(u,w) \le d(u,v) + d(v,w) \le 3 + 1 = 4$, i.e., $d(u,w) = d(u,v) + d(v,w)$, a contradiction. $\square$

Lemma 4.4 ([5]). *Let G be a connected graph and let $v \in V(G)$. If $v \notin I[\mathrm{Ct}(G)]$, then $e(v) \le \mathrm{diam}(G) - 2$.*

Proof. Let $v \in V(G)$ such that $e(v) \ge \mathrm{diam}(G) - 1$. If $e(c) = \mathrm{diam}(G)$, then $v \in \mathrm{Per}(G) \subseteq \mathrm{Ct}(G)$. If $e(c) = \mathrm{diam}(G) - 1$ and $v \notin \mathrm{Ct}(G)$, take a vertex $w \in N(v)$ such that $e(w) = \mathrm{diam}(G)$ and a vertex $u \in \mathrm{Ecc}(w)$. Hence, according to Lemma 4.2, there exists a $w - u$ geodesic containing v. As $e(u) = \mathrm{diam}(G)$, we conclude that $v \in I[w,u] \subseteq I[\mathrm{Per}(G)] \subseteq I[\mathrm{Ct}(G)]$. $\square$

Lemma 4.5 ([5]). *Let G be a bipartite graph of diameter $\mathrm{diam}(G) = d$ and let $v \in V(G)$. If $v \notin I[\mathrm{Ct}(G)]$, then $e(v) \le d - 3$.*

Proof. Let $v \in V(G)$ such that $e(v) \ge d - 2$. If $e(v) \ge d - 1$, then, by Lemma 4.4, $v \in I[\mathrm{Ct}(G)]$. Assume thus that $e(v) = d - 2$ and $v \notin \mathrm{Ct}(G)$. Take a vertex $v_1 \in N(v)$ such that $e(v_1) = d - 1$ and a vertex $z \in \mathrm{Ecc}(v_1)$. Notice that $e(z) \ge d - 1$.

Suppose first that $v_1 \in \mathrm{Ct}(G)$. According to Lemma 4.2, there exists a $v_1 - z$ geodesic containing v. If $z \in \mathrm{Ct}(G)$, then $v \in I[v_1,z] \subseteq I[\mathrm{Ct}(G)]$. Assume thus that $z \notin \mathrm{Ct}(G)$, and take a vertex $z' \in N(z)$ such that $e(z') = d$. Observe that $d(v,z') = d - 3$, as G is bipartite, $d(v,z) = d - 2$, $e(v) = d - 2$ and z and z' are neighbors. In consequence, $d(v_1,z') = d - 2$ and $v \in I[v_1,z'] \subseteq I[\mathrm{Ct}(G)]$.

Assume finally that $v_1 \notin \mathrm{Ct}(G)$. Take a vertex $v_2 \in N(v_1)$ such that $e(v_2) = d$ and a vertex $z \in \mathrm{Ecc}(v_2)$. According to Lemma 4.2, there exists a $v_2 - z$ geodesic containing both v_1 and v. As $e(v_2) = e(z) = d$, we conclude that $v \in I[v_2,z] \subseteq I[\mathrm{Per}(G)] \subseteq I[\mathrm{Ct}(G)]$. $\square$

Proposition 4.7 ([5]). *The contour $\mathrm{Ct}(G)$ of every bipartite graph G of diameter $\mathrm{diam}(G) = d \le 6$ is a geodetic set.*

Proof. Let G be a graph of diameter $\mathrm{diam}(G) = d \le 6$ and $\mathrm{rad}(G) = r$. Observe that if $d < 2r$, then, in all possible cases, $e(v) \ge d - 2$, for every $v \in V(G)$, which according to Lemma 4.4, means that the contour of G is geodetic. Assume thus that $d = 6$ and $r = 3$ and u is a vertex such that $u \notin \mathrm{Ct}(G)$. If $e(u) > 3$, then, according to Lemma 4.4, $u \in I[\mathrm{Ct}(G)]$. Hence we may suppose that $e(u) = 3$ and $u \notin \mathrm{Ct}(G)$.

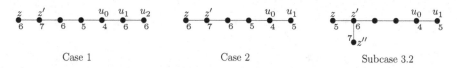

Fig. 4.6 In all cases, $e(u_0) = 4$. In Case 1, $u_0 \in I[z',u_2]$, in Case 2, $u_0 \in I[z',u_1]$ and Subcase 3.2, $u_0 \in I[z'',u_1]$

Notice that for every peripheral vertex z, $d(u,z) = 3$, since $e(u) = 3$ and $e(z) = 6$. Let z_1, z_2 be a pair of antipodal vertices of G, thus satisfying $d(z_1,z_2) = 6$. Hence, if (z_i, y_i, x_i, u) is a $z_i - u$ geodesic, then the path $(z_1, y_1, x_1, u, x_2, y_2, z_2)$ is a $z_1 - z_2$ geodesic containing u, which means that $u \in I[z_1, z_2] \subseteq I[\text{Per}(G)] \subseteq I[\text{Ct}(G)]$. □

Theorem 4.6 ([5]). *The contour* $\text{Ct}(G)$ *of every bipartite graph* G *of diameter at most 7 is a geodetic set.*

Proof. Let G be a graph of diameter $\text{diam}(G) = d \leq 7$. If $d \leq 6$, the according to Proposition 4.7, the contour of G is geodetic. Assume thus that $d = 7$ and u_0 is a vertex such that $u_0 \notin \text{Ct}(G)$. Note that $4 \leq e(u_0) \leq 7$. If $e(u_0) \geq 5$, then, according to Lemma 4.4, $u_0 \in I[\text{Ct}(G)]$. Hence we may suppose that $e(u_0) = 4$.

Let $(u_0, \ldots, u_t)$ a geodesic such that $u_t \in \text{Ct}(G)$ and $e(u_i) = e(u_{i-1}) + 1$, for $1 \leq i \leq t$. Let $z \in \text{Ecc}(u_t)$. Then, according to Lemma 4.2, there exists a $u_t - z$ geodesic containing $(u_0, \ldots, u_t)$. If $z \in \text{Ct}(G)$, then $u_0 \in I[z, u_t] \subseteq I[\text{Ct}(G)]$. Assume that $z \notin \text{Ct}(G)$. Notice that in this case, $5 \leq e(u_t) \leq 6$, since if $e(u_t) = 7$, then $e(z) = 7$, i.e., $z \in \text{Per}(G) \subseteq \text{Ct}(G)$. We distinguish cases.

Case 1: If $e(u_t) = 6$, take a vertex $z' \in N(z)$ such that $e(z') = 7$ (see Fig. 4.6). Notice that $t = 2$ and $d(u_0, z') = 3$, as G is bipartite, $d(u_0, z) = 4$, $e(v) = 4$, and z and z' are neighbors. In consequence, $d(u_2, z') = 5$ and $u_0 \in I[u_2, z'] \subseteq I[\text{Ct}(G)]$.

Case 2: If $e(u_t) = 5$ and $e(z) = 6$, take a vertex $z' \in N(z)$ such that $e(z') = 7$ (see Fig. 4.6). Notice that $t = 1$ and $d(u_0, z') = 3$, as G is bipartite, $d(u_0, z) = 4$, $e(v) = 4$, and z and z' are neighbors. In consequence, $d(u_1, z') = 4$ and $u_0 \in I[u_1, z'] \subseteq I[\text{Ct}(G)]$.

Case 3: If $e(u_t) = 5$ and $e(z) = 5$, take a vertex $z' \in N(z)$ such that $e(z') = 6$. Notice that $t = 1$ and $d(u_0, z') = 3$, as G is bipartite, $d(u_0, z) = 4$, $e(v) = 4$, and z and z' are neighbors. We distinguish cases.

  Subcase 3.1: If $z' \in \text{Ct}(G)$, then $u_0 \in I[u_1, z'] \subseteq I[\text{Ct}(G)]$.

  Subcase 3.2: If $z' \notin \text{Ct}(G)$, then take a vertex $z'' \in N(z')$ such that $e(z'') = 7$ (see Fig. 4.6). Notice that $d(u_0, z'') = 4$, since G is bipartite, $d(u_0, z') = 3$, $e(z'') = 7$, and z and z' are neighbors. In consequence, $d(u_1, z'') = 5$ and $u_0 \in I[u_1, z''] \subseteq I[\text{Ct}(G)]$. □

Theorem 4.7 ([5]). *For every integer* $k \geq 5$ *(resp.* $k \geq 7$*), there is a planar (resp. bipartite) graph* G *of diameter* k *such that its contour* $\text{Ct}(G)$ *is not geodetic.*

Proof. The graph G displayed in Fig. 4.7 is a planar graph such that $\text{diam}(G) = k \geq 5$, $\text{Ct}(G) = \{u, x_0, x_{k-2}\}$, and $I[\text{Ct}(G)] = V(G) - z$.

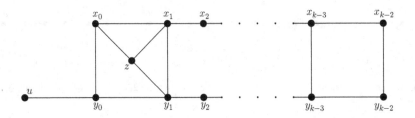

Fig. 4.7 A bipartite graph of diameter 8 whose contour is not geodetic

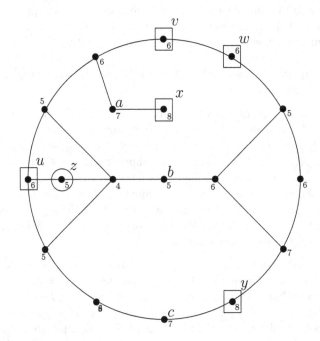

Fig. 4.8 A planar graph of diameter k whose contour is not geodetic

To prove the second statement consider the bipartite graph H displayed in Fig. 4.8, which satisfies $\mathrm{diam}(G) = 8$, $\mathrm{Ct}(G) = \{u,v,w,x,y\}$, and $I[\mathrm{Ct}(G)] = V(G) - z$. Let Ω_1 the graph obtained from H by replacing each vertex of the set $\{v,a,b,c\}$ by a copy of the path P_{s+1}. Check that $\mathrm{diam}(\Omega_1) = 8 + 2s$ and $I[\mathrm{Ct}(\Omega_1)] = V(\Omega_1) - z$. Let Ω_2 the graph obtained from H by replacing each vertex of the set $\{v,b,c\}$ by a copy of the path P_{s+1} and the vertex a by a copy of P_s. Check that $\mathrm{diam}(\Omega_1) = 7 + 2s$ and $I[\mathrm{Ct}(\Omega_2)] = V(\Omega_2) - z$.   □

It remains an *open problem* to characterize (or at least to find strong sufficient conditions) the family of bipartite whose contour is geodetic.

Further results on boundary-type sets related to geodesic convexity in graphs can be found in [5, 23–25, 117, 118].

Chapter 5
Steiner Trees

For a nonempty set W of vertices in a graph G, a connected subgraph of G with the minimum number of edges that contains all of W clearly must be a tree; such a tree is called a *Steiner W-tree*. The *Steiner distance* $d(W)$ of W is the size of a Steiner W-tree [56].

Theorem 5.1 ([136]). *Let $W = \{w_1, \ldots, w_h\}$ be a set of $h \geq 2$ vertices of a graph G. If x is a vertex belonging to the set $\bigcap_{1 \leq i < j \leq h} I[w_i, w_j]$, then $d(W) = \sum_{i=1}^{h} d(w_i, x)$.*

Proof. Suppose that $\cap_{1 \leq i < j \leq h} I[w_i, w_j] \neq \emptyset$ and take a vertex x belonging to this set. We proceed by induction on $h \geq 2$. If $h = 2$, then the result follows from the definition of the geodetic interval between two vertices. Suppose now that $h > 2$ and that the result holds for all $k < h$, $2 \leq k$. Let ρ_i be a shortest $w_i - x$ path, for $1 \leq i \leq h$. Observe that if $i \neq j$, then ρ_i and ρ_j intersect in exactly one vertex, namely x. Let H be the subgraph of G whose edges and vertices are precisely those of $\rho_1, \rho_2, \ldots, \rho_h$. Then, H is a tree that contains the vertices of W. Notice that the size of H is $q(H) = \sum_{i=1}^{h} d(w_i, x)$. If x is one of the vertices of W, say w_h, then $q(H) = \sum_{i=1}^{h-1} d(w_i, x)$. Since $x = w_h \in \cap_{1 \leq i < j \leq h-1} I[w_i, w_j]$, it follows from the induction hypothesis that $d(W - w_h) = \sum_{i=1}^{h-1} d(w_i, x)$. Since $d(W) \geq d(W - w_h)$, the result follows in this case. So we may assume that $x \notin W$. Let T be a Steiner W-tree. We now show that $q(H) \leq q(T)$. Perform a depth first search (DFS) on T starting at one of its end-vertices. Let $w_{i_1}, w_{i_2}, \ldots, w_{i_h}$ be the order in which the vertices of W are visited in the DFS. Observe that when the DFS is complete all the edges of T have been traversed exactly twice. Notice also that $d_T(w_{i_j}, w_{i_{j+1}}) \geq d_H(w_{i_j}, w_{i_{j+1}})$, where subscripts are expressed modulo h. Hence,

$$2q(T) = \sum_{j=1}^{h-1} d_T(w_{i_j}, w_{i_{j+1}}) + d_T(w_{i_h}, w_{i_1})$$

$$\geq \sum_{j=1}^{h-1} d_H(w_{i_j}, w_{i_{j+1}}) + d_H(w_{i_h}, w_{i_1}) = 2q(H).$$

I.M. Pelayo, *Geodesic Convexity in Graphs*, SpringerBriefs in Mathematics,
DOI 10.1007/978-1-4614-8699-2_5, © Ignacio M. Pelayo 2013

The last equality follows from the fact that one can perform a DFS of H by beginning with w_{i_1} and traversing the vertices in the such a way that the vertices of W are visited in the same order as in the DFS of T. The result now follows.   □

The *Steiner interval* $S[W]$ of W consists of all vertices that lie on some Steiner W-tree [136]. If $S[W] = V(G)$, then W is called a *Steiner set* for G.

Lemma 5.1 ([55,116]). *Let $G = (V,E)$ be a connected graph. If $W \subseteq V$ is either a geodetic set or a Steiner set, then* $\text{Ext}(G) \subseteq W$.

Proof. Assume that $v \in \text{Ext}(G)$ such that $v \notin W$. This means that $[W]_g \subseteq V - v$, since the set $V - v$ is convex. Hence, W is neither a hull set nor a geodetic set. Suppose that W is a Steiner set of G. This means that there exists a Steiner W-tree T such that $v \in V(T)$. Certainly, $\deg_T(v) = r \geq 2$, since every leaf of T is in W. If $N_T(v) = \{a_1, \ldots, a_r\}$, then for every $i \in \{1, \ldots, r-1\}$, $a_i a_{i+1} \in E$, since $N(v)$ is a clique of G. In consequence, the subgraph $T' = (T - v) + \{a_1 a_2, \ldots, a_{r-1} a_r\}$ is a tree satisfying both $W \subseteq V(T')$ and $|V(T')| < |V(T)|$, contradicting the fact that T is a Steiner W-tree.   □

The *Steiner number $s(G)$* of G is defined as the minimum cardinality of a Steiner set of G [55].

Theorem 5.2 ([55]). *Let G be a connected graph of order $n \geq 3$. Then,*

1. *$s(G) = 2$ if and only if $g(G) = 2$.*
2. *$s(G) = n$ if and only if $G \cong K_n$.*
3. *If G is not complete, then $s(G) \leq n - \kappa(G)$.*
4. *$s(G) = n - 1$ if and only if G contains a cut-vertex of degree $n - 1$.*

Sketch of proof. Items 1 and 2 are straightforward consequences of the definitions.

To prove item 3, take a non-complete connected graph G s.t. $\kappa(G) = k$. Let $U = \{u_1, \ldots, u_k\}$ be a minimum cutset of G and let $U = \{G_1, \ldots, G_r\}$ be the components of $H \cong G - U$. Then, every vertex u_i $(1 \leq i \leq k)$ is adjacent to at least one vertex of G_j for every j $(1 \leq j \leq r)$. Therefore, every vertex u_i belongs to a Steiner $V(H)$-tree, implying that $V(H)$ is a Steiner set of G, and, consequently, $s(G) \leq |V(H)| = n - \kappa(G)$.

To prove item 4, assume first that G contains a cut-vertex v of degree $n - 1$. By item 3, $s(G) \leq n - 1$. Let W be a minimum Steiner set of G. Notice that $v \in W$ and W contains at least one vertex in each component of $G - v$. Thus every Steiner W-tree of G is star centered at v whose end elements are elements of W and so $S[W] = W \cup \{u\}$. Since $S[W] = V(G)$, it follows that $W = V(G)) \setminus \{v\}$ and $s(G) = n - 1$.

Conversely, suppose that $s(G) = n - 1$, which means that G contains a unique cut-vertex v and $W = V(G) \setminus \{v\}$ is the unique minimum Steiner set of G. To prove that $\deg(v) = n - 1$, assume, to the contrary, that there exists a vertex u such that $vu \notin E(G)$. Let G_1 be the component of $G - v$ containing u and consider the sets $X_1 = N_G(v) \cap V(G_1)$ and $Y_1 = V(G_1) \setminus X_1$. At this point distinguish between two cases, namely, whether or not there exists a vertex of X_1 that is adjacent to no vertex of Y_1, arriving in each case at a contradiction.   □

Table 5.1 Geodetic, hull, and Steiner numbers of some basic graph families

G^a	P_n	C_{2l}	C_{2l+1}	T_n	K_n	$K_{p,q}^{\ b}$	$W_n^{\ c}$	Q_n
$h(G)$	2	2	3	$|Ext(T_n)|$	n	2	$\lfloor \frac{n}{2} \rfloor$	2
$g(G)$	2	2	3	$|Ext(T_n)|$	n	$\min\{4,p\}$	$\lfloor \frac{n}{2} \rfloor$	2
$s(G)$	2	2	3	$|Ext(T_n)|$	n	p	$n-3$	2

$^a G \not\cong K_1$
$^b 2 \leq p \leq q$
$^c n \geq 5$

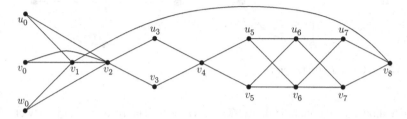

Fig. 5.1 Graph $G(4,6,5)$. Note that $\{u_0,v_0,w_0,v_4,v_8\}$ is a minimum geodetic set and $\{u_0,v_0,w_0,v_1,u_5,v_5\}$ is a minimum Steiner set

Conjecture 5.1. For every graph G of order n and diameter d, $s(G) \leq n-d+1$. Alternatively, find a graph G of order n and diameter 4 such that $s(G) = n-2$.

Theorem 5.3 ([172]). *Let a,b,r be positive integers such that $3 \leq b \leq a$ and $3 \leq r$. Then, there exists a graph G such that $3 \leq r = \mathrm{rad}(G) = \mathrm{diam}(G)$, $g(G) = a$, and $s(G) = b$.*

Sketch of proof. Take a family $\{V_i\}_{i=0}^{2r}$ of $2r+1$ sets such that $|V_0| = b-2$, $|V_1| = |V_2| = |V_r| = |V_{2r}| = 1$, and $|V_i| = a-b+1$ for every $i \in \{3,4,\ldots,r-1,r+1\ldots,2r-1\}$. Consider the graph $G = G(r,a,b) = (V,E)$, where $V = \bigcup_{i=0}^{2r} V_i$ and $E = \{xy : x \in V_i, y \in V_j, 1 \leq i,j \leq 2r, |i-j| \in \{1,2r-1\}\} \cup \{uv : u \in V_0, v \in V_1 \cup V_2\}$. Notice that $\mathrm{diam}(G) = \mathrm{rad}(G) = 3$ (see Fig. 5.1).

Check that $W = V_0 \cup V_1 \cup V_{2r}$ is a minimum geodetic set of G, since $Ext(G) = V_0$ and no set of the form $V_0 \cup \{w\}$ is geodetic. Finally, check that $S = V_0 \cup V_1 \cup V_{r+1}$ is a minimum Steiner set of G, having into account, among other facts, that $V_1 \cup V_2$ is a minimum cutset of G. As $|W| = b$ and $|S| = a$, it follows that $g(G) = b$ and $s(G) = a$, as desired. $\qquad\square$

Conjecture 5.2 ([172]). Let a,b,r be positive integers such that $3 \leq a < b$ and $3 \leq r$. Then, there exists a graph G such that $3 \leq r = \mathrm{rad}(G) = \mathrm{diam}(G)$, $g(G) = a$, and $s(G) = b$.

As was first pointed out in [155], not every Steiner set is geodetic. An example that illustrates this is the graph J_7 shown in Fig. 5.2a. Certainly, the set $W = \{u,v,w\}$ is a minimum Steiner set, but it is not geodetic since $I[W] = V(J_7) - y$. It is also clear that the set $\{a,y,v\}$ is both a minimum Steiner set and a minimum geodetic set,

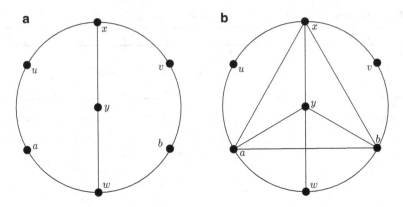

Fig. 5.2 (a) Graph J_7, (b) Graph T_7

which allows us to conclude that $g(G) = s(G) = 3$. Finally, one can quickly check that the graph J_7' obtained from J_7 by adding the edge ab satisfies $s(J_7') = 3$, since $\{u, v, w\}$ is a minimum Steiner set, and $g(J_7') = 4$, since no subset of $V(J_7 + ab)$ of cardinality 3 or less is geodetic. Hence, $s(J_7') < g(J_7')$.

Proposition 5.1 ([116]). *For every pair of integers* $\alpha, \beta \geq 3$, *there exists a connected graph* G *such that* $s(G) = \alpha$, $g(G) = \beta$.

Proof. The case $\alpha \geq \beta$ was proved in [55]. Assume thus that $\alpha < \beta$. Consider the graph $J_{7,m}$ obtained from J_7 (see Fig. 5.2a) by blowing up the two-path $x - y$ into m pieces. Certainly, $\{u, v, w\}$ is a minimum Steiner set. It is also easy to check that the set $A = \{a, v\}$ satisfies $I[A] = V(J_{7,m}) \setminus \{y_i\}_{i=1}^m$, $[A]_g = I^2[A] = V(J_{7,m})$. Hence, $h(J_{7,m}) = 2$. Observe that for every $i \in [m]$ and for every $B \subseteq V(J_{7,m}) \setminus \{x_i, y_i\}$, $I[B] \subsetneq V(J_{7,m})$. From this fact, it is easily seen that $C = \{a, v\} \cup \{y_i\}_{i=1}^m$ is a minimum geodetic set and, hence, that $g(J_{7,m}) = m + 2$. Next, consider the graph $J_{7,m}^k$ obtained from $J_{7,m}$ by adding a set $\{v_1, \ldots, v_k\}$ of k leaves (see Fig. 2.2b). As a direct consequence from Lemma 5.1, we conclude that $\{u, w\} \cup \{v_j\}_{j=1}^k$ (resp. $\{a\} \cup \{v_j\}_{j=1}^k \cup \{y_i\}_{i=1}^m$) is a minimum Steiner (resp. geodetic) set. Hence, taking $m = \beta - \alpha + 1$ and $k = \alpha - 2$, we have a graph G satisfying $s(G) = \alpha$, $g(G) = \beta$. □

At this point, it seems appropriate to ask the following question: *Is every Steiner set a hull set?* As in the previous case, this statement is also false, and here is a counterexample. Consider the graph M_{13} shown in Fig. 5.3a. Certainly, $W = \{1, 4, 7\}$ is a minimum Steiner set, but it is neither a geodetic nor a hull set, since $[W]_g = I[W] = V(M_{13}) - \{a, b, c, d\}$. Observe that the equality $[W]_g = I[W]$ is a direct consequence of the fact that the nine-cycle induced by $V(M_{13}) - \{a, b, c, d\}$ is convex. On the other hand, it is easy to check that the set $S = \{3, 8, d\}$ satisfies $I[S] = V(M_{13}) \setminus \{a, 5, 6\}$, $[S]_g = I^2[S] = V(M_{13})$. Moreover, notice that for every $u, v \in V(M_{13})$, there exists a unique $u - v$ geodesic. Hence, $h(M_{13}) = 3$. Nevertheless, the inequality $h(G) \leq s(G)$ is not true in general. To prove this statement, consider

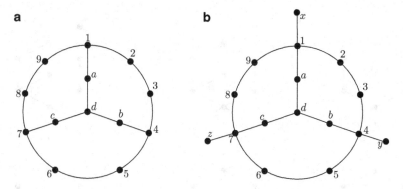

Fig. 5.3 (a) Graph M_{13} s.t. $h(M_{13}) = s(M_{13}) = 3 < 4 = g(M_{13})$. (b) graph M_{16} s.t. $s(M_{16}) = 3 < 4 = g(M_{16}) = h(M_{16})$

the graph M_{16} obtained from M_{13} adding the set of vertices $\Lambda = \{x,y,z\}$ as it is shown in Fig. 5.3b. Certainly, $\text{Ext}(M_{16}) = \Lambda$. Thus, by Lemma 5.1, we obtain that $s(M_{16}) = 3$ and $h(M_{16}) = g(M_{16}) = 4$, since Λ is a Steiner set, $[\Lambda]_g = I[\Lambda] = V(M_{16}) - \{a,b,c,d\}$, and $\{x,y,z,d\}$ is both a geodetic and a hull set.

Theorem 5.4 ([116]). *For every triple a,b,c of integers with $3 \le a \le b \le c$, there exists a connected graph G such that:*

1. $h(G) = a$, $g(G) = b$, $s(G) = c$.
2. $h(G) = a$, $s(G) = b$, $g(G) = c$.
3. $s(G) = a$, $h(G) = b$, $g(G) = c$.

Sketch of proof.

1. The graph in Fig. 5.4b satisfies the equalities $h(G) = k+1$, $g(G) = k+1+p$, $s(G) = k+1+p+m$, for $k \ge 2$, $p \ge 0$, and $m \ge 0$. Hence, if we take $k = a-1$, $p = b-a$, and $m = c-b$, we are done.
2. For $m \ge 1$ and $k \ge 0$, the graph in Fig. 5.4a satisfies $h(G) = m+1$, $g(G) = m+1+k$, $s(G) = m+1+k$. Concretely, the set $W_1 = \{x\} \cup \{v_i\}_{i=1}^m$ is a minimum hull set and $W_2 = W_1 \cup \{u_i\}_{i=1}^k$ is both a minimum geodetic and a minimum Steiner set. As a result, if we take $m = a-1$ and $k = b-a$, we obtain a graph G satisfying $h(G) = a \le b = s(G) = g(G)$.

   Finally, suppose $a \le b < c$. The graph in Fig. 5.4c satisfies the equalities $h(G) = m+2$, $s(G) = m+2+k$, $g(G) = m+2+k+p$, for $m \ge 1$, $k \ge 0$, and $p \ge 1$. If we take $m = a-2$, $k = b-a$, and $p = c-b$, we obtain a graph G such that $h(G) = a \le b = s(G) < g(G) = c$.
3. Consider the graph G constructed with the graphs shown in Fig. 5.5. For example, the graph M_{16} shown in Fig. 5.3b is precisely the graph G for the values $k = 1$, $m = 1$, and $p = 0$. This graph satisfies the equalities $s(G) = k+2$, $h(G) = k+2+m$, and $g(G) = k+2+m+p$, for $k \ge 1$, $m \ge 0$, and $p \ge 0$. If we take $k = a-2$, $m = b-a$, and $p = c-b$, we obtain a graph G such that $3 \le a = s(G) \le h(G) = b \le c = g(G)$.   □

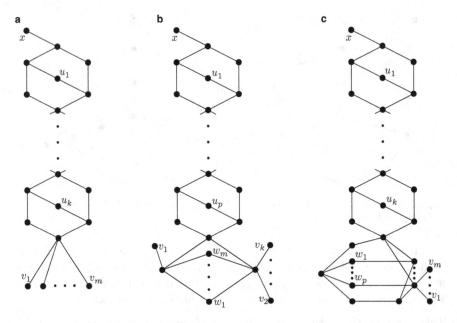

Fig. 5.4 Graphs satisfying $h(G) \leq s(G)$

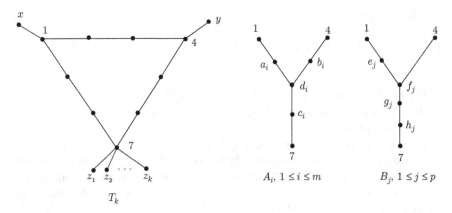

Fig. 5.5 Graph G s.t. $V(G) = V(T_k) \cup \{V(A_i)\}_{i=1}^{m} \cup \{V(B_j)\}_{j=1}^{p}$ and $3 \leq s(G) \leq h(G) \leq g(G)$

Theorem 5.5 ([180]). *If G is a graph of diameter two, then every Steiner set of G is geodetic. In particular, $g(G) \leq s(G)$.*

Proof. Let W be a Steiner set of G and let n be the order of G. If $G[W]$ is connected, then $|W| = n$, and then it is trivially geodetic. Suppose thus that $G[W]$ is not connected. Let $G[B_1], \ldots, G[B_r]$ be the connected components of $G[W]$. Suppose that W is not geodetic and take a vertex $x \notin I[W]$, i.e., such that $x \notin I[u,v]$ for every $u, v \in$

W. Hence, $\emptyset \subseteq N_W(x) \subseteq B_i$, for some $i \in \{1, \ldots, r\}$. Since $\mathrm{diam}(G) = 2$, any Steiner W-tree is formed by r Steiner B_i-trees connected by a nonempty set $\{v_1, \ldots, v_t\} \subseteq V(G) \setminus W$ of vertices, such that, for every $(i, j) \in \{1, \ldots, r\} \times \{1, \ldots, t\}$, $N_W(v_j) \not\subseteq B_i$. Therefore, $x \notin S[W]$, a contradiction. $\square$

Consider the chordal graph shown in Fig. 5.2b. Observe that the vertex set $S = \{u, v, w\}$ is a Steiner set, but it is not geodetic since $I[S] = V(T_7) - y$. Hence, even for the class of chordal graphs the claim—*Every Steiner set is geodetic*—is not true.

A graph G is an *interval graph* if there exists a one-to-one mapping I from $V(G)$ to the set of closed intervals on the real line such that two vertices x, y are adjacent in G if and only if the intervals $I(x)$ and $I(y)$ intersect. The mapping I is called an interval representation of G.

Theorem 5.6 ([116]). *Every Steiner set of an interval graph G is a geodetic set. Hence, every interval graph G satisfies $g(G) \le s(G)$.*

Proof. Let I be an interval representation of G. For each closed interval J in the real line, let $a(J)$ and $b(J)$ denote the left endpoint and the right endpoint of J, respectively.

Let x be a vertex in W such that $I(x)$ has the minimum left endpoint, i.e., $x \in W$ is chosen such that $a(I(x)) = \min\{a(I(w)) : w \in W\}$. Let y be a vertex in W such that $I(y)$ has the maximum right endpoint, i.e., $b(I(y)) = \max\{b(I(w) : w \in W\}$. Suppose first that $x = y$. Then $I(x)$ contains $I(w)$ for all $w \in W$. This means that x is adjacent to all $W - x$, in which case every Steiner tree of W contains only W, implying that $V(G) = W$ and there is nothing to prove.

So we may assume that $x \neq y$. Given any vertex $z \in V(G) - W$, we prove that there exist $w', w'' \in W$ such that z lie on some shortest $w' - w''$ path in G. Since $S[W] = V(G)$ there exists a Steiner tree T of W that contains z. Let P denote the unique path in T between x and y. Let $a^* = a(I(x))$ and $b^* = b(I(y))$. Let $I(P) = \bigcup_{u \in V(P)} I(u)$. Note first that $[a^*, b^*] \subseteq I(P)$. By our choice of x, y, if $v \in W$, then $a(I(v)) \ge a^*$ and $b(I(v)) \le b^*$. Hence $I(v) \subseteq [a^*, b^*] \subseteq I(P)$. In particular, there exists some vertex u on P such that $I(v) \cap I(u) \neq \emptyset$. Thus, if v is not already on P, then it is adjacent to some vertex on P. Let L denote the set of vertices of W not in P. Consider the tree T' obtained from P by adding those vertices of L as leaves to appropriate vertices on P. We have $V(T') = V(P) \cup L$ and $W \subseteq V(T') \subseteq V(T)$. Since T is a Steiner tree of W, we must have $V(T') = V(T)$. Thus, T' is also a Steiner tree of W that contains z. Furthermore, the structure of T' implies that z lies on P.

Now, let w', w'' be vertices of W on P such that the portion of P between w' and w'', denoted $P[w', w'']$, contains z and $V(P[w', w'']) \cap W = \{w', w''\}$; such w', w'' clearly exist. We claim that $P[w', w'']$ is a shortest $w' - w''$ path in G, which will complete our proof. Suppose otherwise that there exists a $w' - w''$ path P' in G shorter than $P[w', w'']$. Consider $P'' = P - P[w', w''] \cup P'$, it is a $x - y$ walk of shorter length than P that contains all of $W \cap V(P)$. Let $I(P'') = \bigcup_{u \in V(P'')} I(u)$. Then $[a^*, b^*] \subseteq I(P'')$. By the same argument as before, each vertex in L is adjacent to some vertex on P''. Thus, $V(P'') \cup L$ induces a connected subgraph of G that contains W and has fewer vertices than T', contradicting T' be a Steiner set of W. This contradiction completes our proof. $\square$

Lemma 5.2 ([116]). *Let $G = (V, E)$ be a connected graph. Let $W \subseteq V$, and let T be a Steiner W-tree in G. Then, $V(T) \subseteq J[W]$.*

Proof. Let H denote the subgraph of G induced by $V(T)$. For each pair $u, v \in W$, let $\rho_{u,v}$ denote an induced u, v-path in H. Note that since H is an induced subgraph of G, $\rho_{u,v}$ is also an induced path in G. In particular, we have $V(\rho_{u,v}) \subseteq J[u, v]$. Let $F = \bigcup_{u,v \in W} \rho_{u,v}$, i.e., F is the union of $\rho_{u,v}$ over all pairs $u, v \in W$. By our discussion above, we have $V(F) = \bigcup_{u,v \in W} V(\rho_{u,v}) \subseteq J[W]$.

Clearly, F is a connected subgraph of H (and hence a connected subgraph of G) that contains all of W. Let F' denote a spanning tree of F, then F' is a subtree in G that contains all of W such that $V(F') \subseteq V(T)$. Since T is a Steiner W-tree, we see that F' must also be a Steiner W-tree and $V(F') = V(T)$. Consequently, we have $V(T) = V(F') = V(F) \subseteq J[W]$. $\qquad\square$

Theorem 5.7 ([116]). *Every Steiner set of a connected graph $G = (V, E)$ is monophonic. In particular, in any distance-hereditary graph, every Steiner set is geodetic.*

Proof. Let $W \subseteq V$ be a Steiner set of G. Then $V(G)$ is the set of all vertices that lie in some Steiner W-tree. By Lemma 5.2, for each Steiner W-tree T in G, we have $V(T) \subseteq J[W]$. Hence, we have $V(G) \subseteq J[W]$. This shows that W is a monophonic set in G, and if G is distance-hereditary, that it is geodetic. $\qquad\square$

Lemma 5.3 ([96]). *Let $G = (V, E)$ be a 3-Steiner distance-hereditary graph. Let $W \subseteq V$, and let T be a Steiner W-tree in G. Then, $V(T) \subseteq I[W]$.*

Sketch of proof. The case $k = 2$ has been proved in Theorem 5.7. Suppose thus that $k \geq 3$. If $|S| = 2$ the result is immediate. Assume hence that $|S| \geq 3$. Let T be a Steiner S-tree and take an arbitrary vertex $v \in V(T) \setminus S$. Notice that if $H = G[V(T)]$, then v is cut-vertex of H, as otherwise $H - v$ is a connected subgraph of G containing S and has smaller order than T, a contradiction. Observe also that each component C of $H - v$ contains at least one vertex of S; otherwise, the removal of $V(C)$ from H produces a connected subgraph of H containing S, of smaller order than H, a contradiction. Take $S' = \{x, y, z\} \subseteq S$ such that y, z do not belong to the same component as x in $H - v$. Let T' be a Steiner S'-tree in H. Then T' is either a path or homeomorphic to $K_{1,3}$. In the latter case the paths beginning at the vertex of degree 3 and terminating at a leaf must be geodesics, as otherwise T' is not a Steiner S'-tree. Let P be the $x - y$ path in T'. Since G is 3-SDH, $|E(T')| = d_H(S') = d_G(S')$. Moreover, v is on P and hence in T', and the $x - v$ path in T' is necessarily a $x - v$-geodesic. If P is a $x - y$ geodesic, we are done. Suppose hence that P is not an $x - y$ geodesic containing v. This means that $d_G(y, z) \geq 2$. At this point, first, check that at least one of $d_G(x, v)$, $d_G(y, v)$, and $d_G(z, v)$ is at least 3 and second, obtain in any case a contradiction. $\qquad\square$

Corollary 5.1. *Every 3-SDH graph G satisfies $g(G) \leq s(G)$.*

Theorem 5.8 ([26]). *For any set of vertices W of an HHD-free graph G, every Steiner W-tree T is contained in the geodesic convex hull of W, i.e., $S[W] \subseteq [W]_g$.*

Sketch of proof. The assertion is proved by induction on the cardinality $|W|$ of W, starting from the obvious case $|W| = 2$. Let T be a tree of S, with $|S| = k \geq 3$. Assume first that there exists a vertex $x \in W$ such that $\deg_T(x) = m \geq 2$. Then, T can be viewed as the union of m subtrees $T_1, \ldots, T_m$ having each of them x as a leaf. Clearly, for every $j \in \{1, \ldots, m\}$, if $W_j = V(T_j) \cap W$, then T_j is a Steiner W_j-tree. Hence, by the inductive hypothesis, for every $j \in \{1, \ldots, m\}$, $V(T_j) \subseteq [W_j]_g$, which means that the desired inclusion $S[W] \subseteq [W]_g$ holds. Suppose thus that $\mathrm{Ext}(T) = W$. We call peripheral vertex of T any vertex z of T with $\deg_T(s) \geq 3$ satisfying the following condition $\mathscr{P}$: there exist two leaves x, y in T such that the only vertex of degree at least 3 belonging to the $x - y$ path P in T is z. We distinguish two cases.

Case 1: There is a peripheral vertex z of T in $[W]_g$. Let P be the $x - y$ path of the above condition $\mathscr{P}$. The path P can be seen as the union of a shortest $x - z$ path P_x and a shortest $y - z$ path, so in this case $V(P) \subseteq [W]_g$. Let T' be the tree obtained from T by removing all vertices from the path P but z. Notice that T' is a Steiner W'-tree, where $W' = W \setminus \{x, y\} \cup \{z\}$. Hence, by the inductive hypothesis, $V(T') \subseteq [W']_g$, which means $V(T) = V(T') \cup V(P) \subseteq [W']_g \cup [W]_g = [W]_g$.

Case 2: All peripheral vertices of T are located outside $[W]_g$. Take a peripheral vertex z of T in $[W]_g$. Let P be the $x - y$ path of $\mathscr{P}$ condition. Notice that P is not a shortest $x - y$ path. Consider the neighbor u_1 of P closer to x in P and the neighbor u_2 of P closer to y in P. Check that $u_1 \neq x$ and $u_2 \neq y$. At this point, using that G is an HHD-free graph, prove first that either z has a neighbor in a $x - y$ geodesic or $u_1 u_2 \in E(G)$ and arrive in both cases, at a contradiction. □

Corollary 5.2. *Every Steiner set of an HHD-free graph G is a hull set. Hence, every HHD-free graph G satisfies $h(G) \leq s(G)$.*

A set $W \subseteq V(G)$ of a graph G is said to be a *Steiner convex* of G if, for every $A \subseteq W$, $S[A] \subseteq W$. Clearly, this definition induces a graph convexity which is called the *Steiner convexity* of G. Similarly to the geodetic case, the smallest Steiner convex set $[W]_s$ containing a set W is called the *Steiner convex hull* of W.

Theorem 5.9 ([26]). *For every set of vertices W of a connected graph G, the following chain of inclusions holds: $[W]_g \subseteq [W]_s \subseteq [W]_m$.*

Proof. It is clear that $[W]_s$ is a g-convex set, since Steiner trees of two-element subsets of W are shortest paths. This gives the first inclusion. For the second one, we will show that $[W]_m$ is a Steiner convex set. Let $A \subseteq [W]_m$ and T be a Steiner tree of A; then, using Lemma 5.2, any vertex in T is in A or in a chordless path between two vertices in A. So, the vertex set of T is contained in $[W]_m$. □

Theorem 5.10 ([26]). *Let G be a connected graph and let $W \subseteq V(G)$. If G is either HHD-free or distance-hereditary, then W is a g-convex set if and only if it is a Steiner convex set.*

Proof. The inclusion $[W]_g \subseteq [W]_s$ in Theorem 5.9 gives the sufficiency. Conversely, suppose that W is a g-convex set and let T a Steiner tree in G of $A \subseteq W$. If G is an HHD-free graph then, according to Theorem 5.8, $V(T) \subseteq [A]_g \subseteq [W]_g = W$, so W is Steiner convex. If G is distance-hereditary, then, according to Lemma 5.2, $V(T) \subseteq J[A] = I[A] \subseteq [W]_g = W$, so W is Steiner convex. $\qquad\square$

Lemma 5.4 ([26]). *Let W be a Steiner convex set of a graph G. Then, a vertex in W is an Steiner extreme vertex of W if and only if it is a simplicial vertex of $G[W]$.*

Proof. If $x \in W$ is a Steiner extreme vertex, then $W \setminus \{x\}$ is Steiner convex and so it is g-convex. Hence, according to Proposition 2.2, x is a simplicial vertex of $G[W]$.

Conversely, let $x \in W$ be a simplicial vertex and $A \subseteq W \setminus \{x\}$. Suppose that x is in a Steiner tree T of A. Using that x is simplicial in W, build a new tree T_0, removing from T edges which are incident to x and adding edges from one of the neighbors of p to all the other ones and then to x. It is clear that T_0 has the same vertex set as T, and p is an end-vertex of T_0. Finally the tree that results by removing x from T_0 contains A and it is smaller than T, a contradiction. So x is not in any Steiner tree of $A \subseteq W \setminus \{x\}$, which means that $W \setminus \{x\}$ is a Steiner convex set and x is a Steiner extreme vertex of W. $\qquad\square$

In [100] (see Corollary 4.6), it was proved that Ptolemaic graphs are the only graphs for which the geodesic convexity is a convex geometry. Next, we show that Ptolemaic graphs are also the only graphs for which the Steiner convexity is a convex geometry.

Theorem 5.11 ([26, 150]). *Let G be a connected graph. Then the following are equivalent:*

1. *G is Ptolemaic.*
2. *The geodesic convexity is a convex geometry.*
3. *The Steiner convexity is a convex geometry.*

Proof. In Corollary 4.6 the equivalence between 1 and 2 was proved. If G is a Ptolemaic graph, then it is distance-hereditary. Hence, according to Theorem 5.10 and Corollary 4.6, the Steiner convexity in G is a convex geometry.

Conversely, let G be a graph such that any Steiner convex set is the Steiner convex hull of its extreme vertices. Let us first see that G has no induced 3-fan. Suppose, to the contrary, that $V(F) = \{a, b, c, d, v\}$ is the vertex set of an induced 3-fan in G (see Fig. 4.5), where $\deg(v) = 4$ and $\deg(a) = \deg(d) = 2$. We consider two cases.

1. If F is Steiner convex, then the extreme vertices of F are a and d, whose Steiner convex hull does not contain neither b nor c, in contradiction with the hypothesis.
2. If F is not Steiner convex, then every extreme vertex of $[F]_s$ belongs to $\{a, d\}$, since the extreme vertices of the convex hull of any set A of vertices must belong to A. In this case, necessarily both a and b must be extreme vertices of $[F]_s$ to satisfy the hypothesis. But $[a, d]_s$ consists of a, d, and their common neighbors and hence does not contain neither b nor c, a contradiction.

Finally, we show that G is chordal using the well-known following characterization of chordal graphs: every induced subgraph has a simplicial vertex. So, let G_0 be an induced subgraph of G and $A = V(G_0)$. There are two cases.

1. If A is a Steiner convex set, then, by hypothesis, A is the Steiner convex hull of its extreme vertices, so A has some extreme vertices which are simplicial vertices of G_0.
2. If A is not Steiner convex, then $S = [A]_s$ is the Steiner convex hull of its extreme vertices, which are simplicial vertices of the subgraph induced by S. Let x be one of these extreme vertices; then $x \in A$ and so x is a simplicial vertex of G_0.   □

Given a pair of vertices u, v of a connected graph $G = (V, E)$, the *edge monophonic closed interval* $J_e[u, v]$ is the set of edges of all monophonic $u - v$ paths. The *edge monophonic closure*, denoted by $J_e[W]$, is defined as the union of all edge closed monophonic intervals over all pairs $u, v \in W$, i.e., $J_e[W] = \bigcup_{u,v \in W} J_e[u, v]$. A vertex set $W \subseteq V$ for which $J_e[W] = E$ is called an *edge monophonic set*. A set $W \subseteq V$ is an *edge Steiner set* if the edges lying in some Steiner W-tree cover E. Notice that:

(1) Every edge Steiner set is a Steiner set.
(2) Every edge geodetic set is geodetic.
(3) Every edge monophonic set is monophonic.
(4) Every edge geodetic set is an edge monophonic set.

It is easy to find examples where the converses of these statements are not true.

Theorem 5.12 ([119]). *Every edge Steiner set of a connected graph is an edge monophonic set. Moreover, in the class of connected interval graphs, every edge Steiner set is an edge geodetic set.*

Further results involving Steiner tree problems related to geodesic convexity in graphs can be found in [15, 18, 26, 28, 47, 55, 96, 115, 116, 136, 150, 152, 155, 172, 180].

Chapter 6
Oriented Graphs

A *directed graph* or *digraph* D is an ordered pair $D = (V,E)$ where $V = V(D)$ is a finite nonempty set of objects called *vertices* and $E = E(V)$ is a set of ordered pairs of distinct vertices of D called *arcs*. If (x,y) is an arc of D, we say that y is an out-neighbor of x and x is an in-neighbor of y. The set of out-neighbors of x is denoted by $N^+(x)$ and the set of in-neighbors of x is denoted by $N^-(x)$. The cardinals $od(x) = |N^+(x)|$, $id(x) = |N^-(x)|$, and $\deg(x) = od(x) + id(x)$ are said to be the *outdegree*, the *indegree*, and the *degree* of x, respectively.

A digraph D is called an *oriented graph* if whenever (x,y) is an arc of D, then (y,x) is not an arc of D. Thus every oriented D graph can be obtained from a graph G by replacing each edge $xy \in E(G)$ with exactly one arc, (x,y) or (y,x). The digraph D is also called an *orientation* of G.

Given two vertices u,v of a digraph D, *a directed $u-v$ path* ρ of length l is any subgraph of D of order $l+1$ such that if $V(\rho) = \{x_i\}_{i=0}^l$, then $x_0 = u$, $x_l = v$, and $E(\rho) = \{(x_i, x_{i+1})\}_{i=0}^{l-1}$. The (directed) distance $d(u,v)$ from u to v in D is the length of a $u-v$ geodesic, i.e., a shortest directed $u-v$ path. The diameter $\mathrm{diam}(D)$ of D is the length of the longest geodesic in G. Although this distance is not a metric, as it clearly lacks the symmetric property, geodetic sets in D and the geodetic number of D can be defined in a natural manner.

Following the terminology used for graphs, given a pair of vertices u and v of a digraph D, we refer to a directed $u-v$ path of length $d(u,v)$ as a $u-v$ geodesic and define $I_D[u,v]$ as the set of all vertices lying on either a $u-v$ geodesic or $v-u$ geodesic of D. For a nonempty subset S of $V(D)$, the geodetic closure I[S] and the convex hull $[S]$ are defined as in the undirected case. As a consequence, a set S of vertices in D is convex if $I[S] = S$, it is a geodetic set if $I[S] = V(D)$, and a hull set if $[S] = V(D)$. The cardinality of a maximum proper convex set of D, a minimum geodetic set of D, and a minimum hull set of D is its convexity number $\mathrm{con}(D)$, its geodetic number $g(D)$, and its hull number $h(D)$, respectively. Clearly, for every nontrivial digraph D, $2 \le h(G) \le g(G) \le n$ and $1 \le \mathrm{con}(D) \le n-1$.

I.M. Pelayo, *Geodesic Convexity in Graphs*, SpringerBriefs in Mathematics, DOI 10.1007/978-1-4614-8699-2_6, © Ignacio M. Pelayo 2013

A vertex v is called an *transitive vertex* of a digraph D if whenever (u,v) and (v,w) are arcs of D, then (u,w) is an arc of D. Notice that every hull set, and thus also every geodetic set, of a digraph must contain all its extreme vertices. An (extreme) vertex is a *source* (resp. a *sink*) if $N^-(v) = \emptyset$ (resp. $N^+(v) = \emptyset$).

Proposition 6.1. *A vertex x is a transitive vertex of a digraph D if and only if it is an extreme vertex, i.e., if $V(D) - x$ is a convex set of D.*

Proof. Assume that x is a transitive vertex of G and take a pair of vertices $u,v \in V(G) - x$. Let ρ a $u-v$ geodesic containing vertex x and take the neighbors $\{a,b\}$ of x belonging to $V(\rho)$. Notice that $d_G(a,b) = 2$, a contradiction since $(a,b) \in E(D)$.

Conversely, let x be a vertex such that $V(D) - x$ is a convex set of D. Suppose that x is neither a source nor a sink, and take a pair of vertices $a \in N^-(x)$ and $b \in N^+(x)$. Since $V(D) - x$ is convex, every vertex lying on some $a - b$ geodesic must belong to $V - x$, which is only true if a and b are adjacent, i.e., if $(a,b) \in E(D)$ □

Corollary 6.1 ([62]). *If D is a oriented graph of order $n \geq 2$, then $\mathrm{con}(D) = n - 1$ if and only if D contains an extreme vertex.*

Lemma 6.1 ([62]). *There is no connected oriented graph of order $n \geq 4$ with convexity number 2.*

Proof. Assume, to the contrary, that there is a connected oriented graph D of order $n \geq 4$ having convexity number 2. Let $\{u,v\}$ be a maximum proper convex set of D. Since $[\{u,v\}] = \{u,v\}$, it follows that D cannot contain both a directed $u - v$ path and a directed $v - u$ path. Assume then that there is no directed $v - u$ path in D. Let $U = \{x : D \text{ contains a directed } x - u \text{ path }\}$. Since D contains no directed $v - u$ path, $v \notin U$ and therefore $|U| \leq n - 1$. If $U = \{u\}$, then u is a source and so $\mathrm{con}(D) = n - 1$ by Corollary 6.1. Hence $|U| \geq 2$.

Next, observe that D has no arc of the form (w,x), where $x \in U$ and $w \in V(D) \setminus U$. Hence, every geodesic connecting two vertices of U uses only vertices in U. Thus, U is convex. Since $\mathrm{con}(D) = 2$ and U is a convex set with $2 \leq |U| \leq n - 1$, it follows that $|U| = 2$. Let $U = \{x,u\}$. Since there is no arc in D with its initial vertex in $V(D) \setminus U$ and its terminal vertex in U, it follows that x is a source and so $\mathrm{con}(D) = n - 1 \geq 3$, a contradiction.   □

Theorem 6.1 ([62]). *For every two integers k and n with $1 \leq k \leq n - 1$, $k \neq 2$, and $n \geq 4$, there exists an oriented graph D of order n such that $\mathrm{con}(D) = k$.*

Proof. If $k = 1$, then the directed cycle of order n has the desired property since $\mathrm{con}(\overrightarrow{C}_n) = 1$. If $k = n - 1$, then the directed path $\overrightarrow{P}_n$ has the desired property. So, we may assume that $3 \leq k \leq n - 2$. Let D be the oriented graph obtained from the directed cycle $\overrightarrow{C}_k$, where $V(\overrightarrow{C}_k) = \{v_i\}_{i=1}^k$ and $E(\overrightarrow{C}_k) = (v_k, v_1) \cup \{(v_i, v_{i+1})\}_{i=1}^{k-1}$, and the directed path $\overrightarrow{P}_{n-k}$, where $V(\overrightarrow{P}_{n-k}) = \{u_i\}_{i=1}^{n-k}$ and $E(\overrightarrow{P}_{n-k}) = \{(u_i, u_{i+1})\}_{i=1}^{n-k-1}$ by adding the two arcs (v_1, u_1) and (u_{n-k}, v_3).

We show that $\mathrm{con}(D) = k$. Since $V(\overrightarrow{C}_k)$ is convex, $\mathrm{con}(D)k \geq$. Suppose that $\mathrm{con}(D) > k$. Notice that, for every pair i,j with $1 \leq i < j \leq n - k$, $[\{u_i, u_j\}] = V(D)$. Hence, is S_0 is a set of vertices of D containing two vertices of U, then $[S_0] = V(D)$. Moreover, $[\{v_i, u_j\}] = V(D)$ for $i = 1, 3$ and $1 \leq j \leq n - k$. Let S be a proper convex set of D with $|S| \geq k+1$. Since S contains at most one vertex of U, it follows that $S = V(\overrightarrow{C}_k) \cup \{u_j\}$, where $1 \leq j \leq n-k$. This implies that $\{v_1, u_j\} \subset S$ and so $[S] = V(D)$, which is a contradiction. Therefore, $con(D) = k$. □

A digraph D is *transitive* if all of its vertices are transitive. A digraph D is *antidirected* if every vertex of D is either a source or a sink. Observe, first, that every antidirected digraph is both oriented and transitive and, second, that a graph G has an antidirected orientation if and only if it is bipartite.

Theorem 6.2 ([50,67]). *If D is a oriented graph of order n, then the following are equivalent: (i) $h(D) = n$, (ii) $g(D) = n$, (iii) D is transitive.*

Proof. Assume first that $h(D) < n$. Then, there exists a vertex v in D such that $S = V(D) - v$ is a hull set of D. This means that v lies on some $u - w$ geodesic of length 2, where $u, v \in S$. However, then, $(u, w) \notin E(D)$ and D is not transitive.

Conversely, assume that D is an oriented graph that is not transitive. Then there exist distinct vertices u, v, and w such that $(u, v), (v, w) \in E(D)$, but $(u, w) \notin E(D)$. Then $S = V(D) - v$ is a geodetic set and so $g(D) < n$. □

Proposition 6.2 ([50, 67]). *For every two integers k and n with $2 \leq k \leq n$, there exists an orientation of P_n such that $h(D) = g(D) = k$.*

Proof. Suppose that $V(P_n) = \{v_1, v_2, \ldots, v_n\}$. We construct an oriented graph D from P_n by directing the two edges incident with v_i towards v_i for all even i with $i < k$. If k is odd, then each edge $v_i v_{i+1}$ with $k \leq i \leq n-1$ is directed as $(v_{i+1} v_i)$. This completes the construction of D. Since the vertices $v_1, \ldots, v_{k-1}$ are a source or a sink, each of these vertices belongs to every hull set of D. Since $\{v_1, \ldots, v_{k-1}\}$ is not a geodetic set, but $\{v_1, \ldots, v_{k-1}, v_n\}$ is, $h(D) = g(D) = k$. If k is even, then we direct each edge $v_i v_{i+1}$ for $k-1 \leq i \leq n-1$ to produce the arc (v_i, v_{i+1}) and to complete the construction of D. Again, each vertex v_j ($1 \leq j \leq k-1$) belongs to every hull set of D, but $\{v_1, \ldots, v_{k-1}\}$ is not a geodetic set. Since $\{v_1, \ldots, v_{k-1}, v_n\}$ is a geodetic set, $h(D) = g(D) = k$. □

Proposition 6.3 ([50]). *For every two integers k and n with $2 \leq k \leq n$, there exists an orientation of K_n, i.e., a tournament, such that $h(D) = g(D) = k$.*

Proof. Suppose that $V(K_n) = \{v_1, v_2, \ldots, v_n\}$. We consider the tournament T defined as follows: for $1 \leq i \leq n-k+1$, let $(v_i, v_{i+1}) \in E(T)$; otherwise, for $1 \leq i < j \leq n$, let $(v_j, v_i) \in E(T)$. Since $d_T(v_1, v_{n-k+2}) = n-k+1$, $I[v_1, v_{n-k+2}] = \{v_1, \ldots, v_{n-k+2}\}$. For $n-k+3 \leq i \leq n$, the vertex v_i lies on a geodesic if and only if it is the initial or terminal vertex of a geodesic (of length 1). Hence $\{v_{n-k+3}, \ldots, v_n\}$ is a subset of every hull set, and thus also every geodetic set, in T. Since $\{v_1, v_{n-k+2}, v_{n-k+3}, \ldots, v_n\}$ is a hull set, and thus also a geodetic set, we obtain that $h(T) = g(T) = k$. □

Corollary 6.2 ([50,67]). *For every two integers k and n with $2 \leq k \leq n$, there exists an oriented graph D of order n such that $h(D) = g(D) = k$.*

Conjecture 6.1. For every three integers h, k, and n with $2 \leq h \leq k \leq n-1$, there exists an oriented graph D of order n such that $h(D) = h$ and $g(D) = k$.

Fig. 6.1 An oriented graph D
with $h(D) = |\text{Ext}(G)| =$
$|\{u,v\}| = 2$ and
$g(D) = |\{u,v,z\}| = 3$

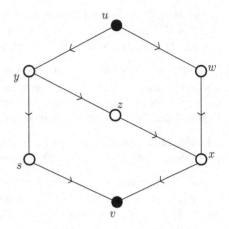

Theorem 6.3 ([50]). *If D is a oriented graph of order n and diameter d, then $g(D) \leq n - d + 1$. Moreover, this upper bound is tight.*

Proof. Let u and v be vertices of D for which $d(u,v) = d$ and let $u = v_0, v_1, \ldots, v_d = v$ be a $u - v$ geodesic. If $S = V(D) \setminus \{v_1, v_1, \ldots, v_{d-1}\}$, then $I[S] = V(D)$ and so $g(D) \leq |S| = n - d + 1$. Moreover, the oriented graph D constructed in the proof of Theorem 6.2 has order n, diameter $d = n - k + 1$, and geodetic number $g(D) = k = n - d + 1$, which proves the sharpness of the upper bound. $\square$

Theorem 6.4 ([67]). *For every pair a, b of integers with $2 \leq a \leq b$, there exists an oriented graph D such that $h(D) = a$ and $g(D) = b$.*

Sketch of proof. Assume first that $a = b \geq 2$. Consider the oriented graph D whose underlying graph is the star $K_{1,a}$, where $V(D) = \{v\} \cup \{v_i\}_{i=1}^a$, $\deg(v) = a - 1$, and $E(D) = (v, v_1) \cup \{(v_i, v)\}_{i=2}^a$. Check that $\text{Ext}(D) = V(D) - v$ is both a minimum hull and a minimum geodetic set.

Suppose next that $2 \leq a < b$. For $1 \leq i \leq b - a$, let D_i be a copy of the oriented graph displayed in Fig. 6.1, where $V(D) = \{u_i, w_i, x_i, v_i, s_i, y_i, z_i\}$. Consider the oriented graph D obtained from the oriented graphs $\{D_1, \ldots, D_{b-a}\}$ by adding (1) a set $\{t_i\}_{i=1}^{a-1}$ of $a - 1$ new vertices, (2) the set $\{(v_{b-a}, t_i)\}_{i=1}^{a-1}$ of new arcs, and (3) the set $\{(u_i, u_{i+1})\}_{i=1}^{b-a-1} \cup \{(v_i, v_{i+1})\}_{i=1}^{b-a-1}$ of new arcs. Check, first, that $\text{Ext}(D) = \{u_1\} \cup \{t_i\}_{i=1}^{a-1}$ is the unique minimum hull set of D and, second, that $\{u_1\} \cup \{t_i\}_{i=1}^{a-1} \cup \{z_i\}_{i=1}^{b-a}$ is a minimum geodetic set. $\square$

We have seen that for every two integers k and n with $2 \leq k \leq n$, there exists an orientation of the path P_n (and also of K_n) having geodetic number k. Given a graph G, the *geodetic spectrum* $S(G)$ is defined as the set of geodetic numbers of all orientations of G, i.e., $S(G) = \{g(D) : D \text{ is an orientation of } G\}$. Hence, we know that $S(P_n) = S(K_n) = \{2, \ldots, n\}$ for every $n \geq 2$. Moreover, in [39] the geodetic spectrum of some other basic families was studied (see Table 6.1).

Table 6.1 Geodetic and convexity spectra of some basic graph families [39, 174, 179]

G^a	P_n	C_n	T_n	$K_n{}^b$	$K_{p,q}{}^{c\ d}$
$S(G)$	$\{i\}_{i=2}^n$	$\{3\}\cup\{2i\}_{i=1}^{\lfloor\frac{n}{2}\rfloor}$	$\{i\}_{i=|Ext(T_n)|}^n$	$\{i\}_{i=2}^n$	$\{i\}_{i=2}^n$
$S_c(G)$	$\{n-1\}$	$\{1,n-1\}$	$\{n-1\}$	$\{i\}_{i=1}^{n-1}\setminus\{2,4\}$	$\{1,n-1\}\cup\{\lceil\frac{n}{2}\rceil+i\}_{i=0}^{\lfloor\frac{n}{2}\rfloor-4}$

$^aG\not\cong K_1$
^{b}If $n\in\{5,6\}$, then $4\in S_c(K_n)$
$^c 2\le p\le q$
^{d}If $2\le p\le 3$, then $S_c(K_{p,q})=\{1,n-1\}$

Table 6.2 Orientable parameters of some basic graph families [50, 62, 67]

G^a	P_n	C_{2l}	C_{2l+1}	T_n	K_n	$K_{p,q}{}^b$
$h^-(G)$	2	2	2	$|Ext(T_n)|$	2	2
$g^-(G)$	2	2	2	$|Ext(T_n)|$	2	2
$h^+(G)$	n	n	$n-1$	n	n	n
$g^+(G)$	n	n	$n-1$	n	n	n
$con^-(G)$	$n-1$	1	1	$n-1$	1	1
$con^+(G)$	$n-1$	$n-1$	$n-1$	$n-1$	$n-1$	$n-1$

$^aG\not\cong K_1$
$^b 2\le p\le q$

In general, given an arbitrary graph G, there are orientations of G having distinct geodetic numbers. This fact suggests the following definitions.

For a connected nontrivial graph G, the *lower orientable geodetic number* $g^-(G)$ of G is defined as the minimum geodetic number of an orientation of G and the *upper orientable geodetic number* $g^+(G)$ as the maximum such geodetic number. The *lower orientable hull number* $h^-(G)$ and the *upper orientable hull number* h^+ are similarly defined.

Proposition 6.4 ([50, 67]). *Let G be a nontrivial graph of order n. If $G\in\{T_n,C_n,K_n,K_{r,n-r}\}$, then $h^-(G)=g^-(G)=2$ and $h^+(G)=g^+(G)=n$, with just the following two exceptions: $h^-(T_n)=g^-(T_n)=|Ext(T_n)|$ and $h^+(C_{2l+1})=g^+(C_{2l+1})=n-1$, for every $l\ge 2$ (see Table 6.2).*

Sketch of proof. All of these results are straightforward consequences of Theorem 6.2, Proposition 6.3, and the following facts: (1) Every hull set, and thus also every geodetic set, of a digraph must contain all its extreme vertices. (2) If a graph G contains a Hamiltonian path, then $h^-(G)=g^-(G)=2$. (3) A graph G has an antidirected orientation is and only if it is bipartite. (4) If $r\ge 2$, there exists an orientation of $K_{r,n-r}$ such that if x,y are two adjacent vertices, then x is a source, y is a sink and the rest of vertices are not transitive. □

Theorem 6.5 ([50, 102]). *For every connected graph $G=(V,E)$ of order $n\ge 3$, $2\le g^-(G)<g^+(G)\le n$ and $2\le h^-(G)<h^+(G)\le n$.*

Sketch of proof. If G is the complete graph, then, according to Proposition 6.4, $h^-(G) = g^-(G) = 2$ and $h^+(G) = g^+(G) = n$. Suppose thus that G is not complete. Take three vertices v_0, v_1, v_2 such that $d(v_0, v_2) = 2$ and $v_1 \in N(v_0) \cap N(v_2)$. Denote $U_0 = \{v_0, v_1, v_2\}$, $U = V \setminus U_0$, $N_1 = N(v_1)$, $N_2 = N(v_2)$, and $N_{12} = N(v_1) \cap N(v_2)$. Consider the partition $\{U_i\}_{i=0}^5$ of V where: $U_1 = U \cap N_1$, $U_2 = U \cap N_2$, $U_3 = U \cap N_{12}$, $U_4 = N(U_3) \setminus \cup_{i=0}^3 U_i$, and $U_5 = V \setminus \cup_{i=0}^4 U_i$. Let D_2 be an orientation of G such that (1) v_0, v_2 are sources, (2) v_1 is a sink, and (3) if $xy \in V(G)$, and $(x,y) \in (U_1 \times (U \setminus U_1)) \cup (U_4 \times U_3) \cup ((U \setminus U_2) \times U_2)$, then $(x,y) \in V(D_2)$. Let D_1 be obtained from D_2 by reversing the orientation of the arcs incidents to v_2. Then, prove that if S is a hull set of D_2, then, first, $I_{D_2}[S] \subseteq I_{D_1}[S - v_1]$ and, second, $I_{D_2}^l[S] \subseteq I_{D_1}^l[S - v_1]$, for any $l \geq 2$, proceeding by induction. So, if S is a minimum hull set of D_2, then $I_{D_2}^k[S] = V$, for some k. Hence, $I_{D_1}^k[S - v_1] = V$, i.e., $S - v_1$ is a hull set of D_1. As v_1 is a sink of D_2, $v_1 \in S$, and thus $h^-(G) \leq H(D_1) \leq |S - v_1| < |S| \leq h(D_2) \leq h^+(G)$. The inequality $g^-(G) < g^+(G)$ is similarly proved, just by taking $k = 1$. $\square$

Lemma 6.2 ([67]). *If G is a connected graph with at least 3 vertices and T is a spanning tree of G with k leaves, then $g^-(G) \leq k$.*

Proof. Let S be the set of all leaves of T and let x be a leaf of T. Consider any orientation D of G such that (i) If $uv \in E(T)$ and $d_T(r, u) > d_T(r, v)$, then $(v, u) \in D$. (ii) If $uv \in E(G) \setminus E(T)$ and $d_T(r, u) > d_T(r, v)$, then $(u, v) \in D$. In D, we have that the paths from r to the other leaves in T are geodesics of D, i.e., S is a geodetic set of D. Hence, $g^-(G) \leq g(D) \leq |S| = k$. $\square$

Theorem 6.6 ([122]). *For every connected graph G of order $n \geq 3$, $g^-(G) < h^+(G)$.*

Proof. If either $n = 3$ or G is complete, then the inequality trivially holds. Suppose that G is a non-complete graph of order $n \geq 4$ having a Hamiltonian path. Take two vertices u, v such that $d(u, v) \geq 2$. Take $x \in N(u)$ and $y \in N(v)$. If $x \neq y$, any orientation D_1 of G such that u, y are sources and x, v are sinks satisfies $h^+(G) \geq h(D_1) \geq 4 = h^-(G) + 2$. If $x = y$, any orientation D_2 of G such that u, v are sources and x is a sink satisfies $h^+(G) \geq h(D_2) \geq 4 = h^-(G) + 2$, since $n \geq 4$ and $\{u, x, v\}$ is a convex set.

Assume that G has no Hamiltonian path. Let T be a spanning tree of G obtained by the depth-first search algorithm where $V(T) = \{u_1, u_2, \ldots, u_n\}$ and the leaves of T be $\{v_1, v_2, \ldots, v_k\}$, with $k \geq 3$ as G has no Hamiltonian path. If u_1 is not in T, then $\{v_1, v_2, \ldots, v_k\}$ is an independent set of G. Construct an orientation D_1 of G by $(u_i, u_j) \in E(D_1)$ if and only if $u_i u_j \in E(G)$ and $i < j$. Then u_1 is a source and $v_1, v_2, \ldots, v_k$ are sinks of D_1; i.e., every hull set of D_1 contains $u_1, v_1, v_2, \ldots, v_k$. Hence, by Lemma 6.2, $g^-(G) \leq k < k + 1 \leq h^+(G)$. Assume now that u_1 is a leaf in T. We distinguish two cases.

Case 1: $u_1 v_i \notin E(G)$, for all $i \in \{2, \ldots, k\}$. In this case, $\{v_1, v_2, \ldots, v_k\}$ is an independent set of G and we can construct an orientation D_2 of G such that $v_1, v_2, \ldots, v_k$ be sinks of D_2. Hence, by Lemma 6.2, $g^-(G) \leq k < k + 1 \leq h^+(G)$, since $\{v_1, v_2, \ldots, v_k\}$ is a proper convex set of D_2.

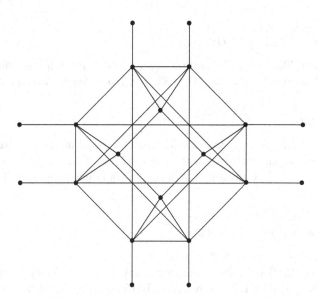

Fig. 6.2 Graph H. Notice that $h(H) = g(H) = 10$

Case 2: $u_1 v_j \in E(G)$, for some $j \in \{2, \ldots, k\}$. $\{v_1, v_2, \ldots, v_k\}$. Let i be the smaller integer such that $\deg_t(u_i) \geq 3$. Take the spanning tree T' of G such that $E(T') - u_{i-1}u_i + u_1 v_j$. As T' has $k - 1$ leaves, it follows, by Lemma 6.2, that $g^-(G) \leq k - 1$. Let D_3 be the orientation of G such that $(u_i, u_j) \in E(D_3)$ if and only if $u_i u_j \in E(G)$ and $i < j$. Then, u_1 is a source and $v_2, \ldots, v_k$ are sinks of D_1, and hence $h^+(G) \geq h(D_3) \geq k$.   □

Corollary 6.3. *For every connected graph G of order $n \geq 3$, $h^-(G) \leq g^-(G) < h^+(G) \leq g^+(G)$.*

Theorem 6.7 ([122]). *For every pair a, b of nonnegative integers, there exists a connected graph G such that $g^-(G) - h^-(G) = a$ and $g^+(G) - h^+(G) = b$.*

Sketch of proof. This result is a direct consequence of the following assertions:

- Let $C(n, t)$ be the graph obtained from the cycle C_n by adding a pair of vertices $\{x, y\}$ and the pair of edges $\{1x, ty\}$, where $V(C_n) = \{i\}_{i=1}^n$. If $n \geq 5$, $t \neq \frac{n}{2} + 1$, and $3 \leq t \leq n - 1$, then $h^-(C(n, t)) = 2$, $g^-(C(n, t)) = 3$, and $h^+(C(n, t)) = g^+(C(n, t)) = n + 1$.
- The graph H depicted in Fig. 6.2 satisfies $h^-(H) = g^-(H) = 10$, $h^+(H) = 17$, and $g^+(H) = 18$.
- Let G_1 and G_2 be a pair of disjoint graphs. Let $x_1 \in V(G_1)$ and $x_2 \in V(G_2)$. If $\deg_{G_1}(x_1) = 1$ and $\deg_{G_2}(x_2) = 1$, then the graph G obtained from G_1 and G_2 by adding the edge $x_1 x_2$ satisfies the following properties:

  – $h^+(G) = h^+(G_1) + h^+(G_2)$.
  – $g^+(G) = g^+(G_1) + g^+(G_2)$.

- $h^-(G) = h^-(G_1) + h^-(G_2) - 2$.
- $g^-(G) = g^-(G_1) + g^-(G_2) - 2$.   □

Theorem 6.8 ([79]). *A graph G satisfies that $g^+(G) \geq g(G)$, if it satisfies at least one of the following conditions: (i) is K_3-free, (ii) is four-colorable, and (iii) is chordal.*

Sketch of proof.

(i) Let $G = (V,E)$ be a K_3-free graph. Take a maximal independent set V_1 of G and then a maximal independent set V_2 of $G - V_1$. Check that $A = V_1 \cup V_2$ is a geodetic set of G. Consider an orientation D of G satisfying the following two conditions:

- All vertices of V_1 are sources.
- All vertices of V_2 are sinks.

Hence, $g^+(G) \geq g(D) \geq |A| \geq g(G)$.

(ii) Let $G = (V,E)$ be a four-colorable graph. Let V_1, V_2, V_3, V_4 be the sets of vertices colored by $1,2,3,4$, respectively. Consider an orientation D of G such that if $i < j$, $v_i \in V_i$, and $v_j \in V_j$, then $(v_i, v_j) \in E(D)$. Notice that D has no directed cycles. Observe also that D satisfies the following property $\mathscr{P}$: for any $u, v \in V(G)$, $I_D[u,v] \cup I_D[v,u] \subseteq I_G[u,v]$.

Check that for every graph G satisfying property $\mathscr{P}$, $g(G) \leq g(D) \leq g^+(G)$.

(iii) Let $G = (V,E)$ be a chordal graph of order n. Let $v_1, v_2, \ldots, v_n$ a simplicial elimination ordering of V. Let S be a maximum independent set of G. Consider the sets $U = \{v \in V \setminus S : |N_G(v) \cap S| \geq 2\}$ and $W = V \setminus (S \cup U)$.

Let D be an orientation of G satisfying the following two conditions:

- If $x \in S$ and $y \in V \setminus S$, then $(x,y) \in E(D)$.
- If $x \in U$ and $y \in V \setminus W$, then $(x,y) \in E(D)$.
- If $v_i, v_j \in U$, $v_i v_j \in E(G)$ and $i < j$, then $(v_i, v_j) \in E(D)$.
- If $v_i, v_j \in W$, $v_i v_j \in E(G)$ and $i <_* j$, then $(v_i, v_j) \in E(D)$.

Check that, if G is not complete, then every minimum geodetic set of D is a geodetic set of G.   □

Conjecture 6.2 ([139]). Every graph G satisfies that $g^+(G) \geq g(G)$.

Theorem 6.9 ([39, 50, 67]). *For every two integers n and m with $1 \leq n-1 \leq m \leq \binom{n}{2}$, there exists a connected graph G of order n and size m such that, for each integer k with $2 \leq k \leq n$, there exists an orientation D of G with $h(D) = g(D) = k$.*

Proof. We prove by induction on n the following more general statement: For every two integers n and m with $1 \leq n-1 \leq m \leq \binom{n}{2}$, there exists a connected graph G of order n and size m having a vertex v such that (a) G contains a Hamiltonian path with initial vertex v and (b) for each $3 \leq k \leq n$, there exists an orientation D of G having v as a source and $h(D) = g(D) = k$.

Since the statement is certainly true if $n \leq 3$, we assume that $n \geq 4$. Suppose that the statement is true for $n - 1$. We consider two cases.

Case 1: $2n - 3 \leq m \leq \binom{n}{2}$. Then, $(n-1) - 1 \leq m - (n-1) \leq \binom{n-1}{2}$. By the induction hypothesis, there exists a connected graph G' of order $n-1$ and size $m - (n-1)$ having a vertex v' such that G' contains a Hamiltonian path starting at v' and for each $3 \leq k \leq n-1$, there exists an orientation D' of G' having v' as a source and $h(D') = g(D') = k$. Consider the graph $G = G' \vee K_1$, with $V(K_1) = \{v\}$. Since G' contains a Hamiltonian path starting at v', it follows that G contains a Hamiltonian path starting at v. Suppose next that $3 \leq k \leq n$, i.e., $2 \leq k-1 \leq n-1$. By the induction hypothesis, there exists an orientation D' of G' having v' as a source and $h(D) = g(D) = k - 1$. We extend the orientation D' of G' to an orientation D of G by directing each edge incident with v in G away from v, i.e., in such a way that v be a source of D. It is a routine exercise to check that $h(D) = g(D) = g(D') + 1 = h(D') + 1 = k$.

Case 2: $n - 1 \leq m \leq 2n - 4$. Then, $(n-1) - 1 \leq m - 1 \leq 2n - 5$. By the induction hypothesis, there exists a connected graph G' of order $n-1$ and size $m-1$ having a vertex v' such that G' contains a Hamiltonian path starting at v' and for each $3 \leq k \leq n-1$, there exists an orientation D' of G' having v' as a source and $h(D') = g(D') = k$. Consider the graph G obtained from G by adding a new vertex v and joining it to v'. Since G' contains a Hamiltonian path starting at v', it follows that G contains a Hamiltonian path starting at v. Suppose next that $3 \leq k \leq n$. We consider two cases.

Subcase 2.1: $3 \leq k \leq n-1$. By the induction hypothesis, there exists an orientation D' of G' having v' as a source and $h(D') = g(D') = k$. We extend the orientation D' of G' to an orientation D of G by directing the edge vv' as (v, v'), i.e., in such a way that v be a source of D. It is a routine exercise to check that $h(D) = g(D) = g(D') = h(D') = k$.

Subcase 2.2: $k = n$. By the induction hypothesis, there exists an orientation D' of G' having v' as a source and $h(D') = g(D') = n - 1$. Consider the oriented graph D^* obtained from D' by reversing the direction of all of its edges. We extend the orientation D^* of G' to an orientation D of G by directing the edge vv' as (v, v'), i.e., in such a way that v be a source and v' be a sink of D. It is a routine exercise to check that $h(D) = g(D) = g(D^*) + 1 = h(D^*) + 1 = n$. $\quad\square$

For a connected nontrivial graph G, the *lower orientable convexity number* $\mathrm{con}^-(G)$ of G is defined as the minimum convexity number among the orientations of G and the *upper orientable convexity number* $\mathrm{con}^+(G)$ as the maximum such convexity number.

For any nontrivial connected graph G we can make an arbitrary vertex v extreme by orienting all edges of G that are incident with v away from v and direct all other edges arbitrarily, producing an orientation D of G with a source. Then it follows from Corollary 6.1 that $\mathrm{con}^+(G) = n - 1$. Hence, for every connected nontrivial graph G, $1 \leq \mathrm{con}^-(G) \leq \mathrm{con}^+(G) = n - 1$. In Table 6.2, both parameters for some basic graph families are displayed (see [62] for more details). Next, we characterize those graphs for which $\mathrm{con}^-(G) < \mathrm{con}^+(G)$.

Theorem 6.10 ([102]). *Every graph G with $\delta(G) \geq 2$ can be oriented so that all its vertices are non-extreme. Thus, for a connected graph G with at least 3 vertices, $\text{con}^-(G) < \text{con}^+(G)$ if and only if G has no end-vertices.*

Proof. Since G has minimum degree at least 2, it contains a cycle. Find a maximal set M of edge-disjoint chordless cycles, and orient their edges to make them directed cycles. We call oriented vertex every vertex incident to some oriented edge. We claim that every oriented vertex v is now non-extreme. If v is on a triangle uvw in M, then (u,v), (v,w), and (v,w) are all arcs. Otherwise, v is on a chordless cycle in M of length at least 4, with neighbors, say, u and w, where $uw \notin E(G)$, but (u,v) and (v,w) are arcs.

We now show that if there are unoriented vertices, we can orient one or more while maintaining the property that all oriented vertices are non-extreme. Any unoriented vertex u must be on a path $u_0, \ldots, u, \ldots, u_{r+1}$ joining two distinct oriented vertices u_0 and u_{r+1}, since $\delta(G) \geq 2$ and the set M of edge-disjoint cycles was chosen to be maximal. Taking r to be as small as possible ensures that the internal vertices $u_1, \ldots, u_r$ are all unoriented. Directing the path as $(u_0, u_1), \ldots, (u_r, u_{r+1})$ ensures that $u_1, \ldots, u_r$ all have positive indegree and outdegree. Moreover, if $r > 1$, then, for $1 \leq i \leq r$, $u_{i-1}u_{i+1} \notin E(G)$, and thus u_i is non-extreme. If $r = 1$, then we might have to orient differently if $u_0 u_2$ is an edge of G. Since $u_0 u_1$ and $u_1 u_2$ are unoriented, and M is maximal, $u_0 u_2$ must be oriented, say as (u_0, u_2). Now orienting $u_0 u_1$ and $u_1 u_2$ as (u_1, u_0) and (u_2, u_1) ensures that u_1 is on a directed triangle and is thus non-extreme.   □

Further results related to geodesic convexity in directed graphs can be found in [39, 50, 62, 67, 102, 107, 122, 174, 179].

Chapter 7
Computational Complexity

Let G be a graph of order n and size m and S a proper subset of $V(G)$. The geodetic closure $I[S]$ can be obtained in $O(|S|m)$ time by applying $|S|$ times Bread First Search, each one starting from a distinct $s \in S$. In particular, the problem of verifying if S is convex can be done in $O(|S|m)$ time. By iteratively applying this process, the convex hull $[S]$ can be determined in $O([S]m)$ time [81].

Given a graph G and a positive integer k, the problem of deciding whether $g(G) \leq k$ is called the *Geodetic Number Problem*.

Theorem 7.1 ([80,84]). *The Geodetic Number Problem restricted to either bipartite or chordal graphs is NP-complete.*

Proof. Suppose first that $G = (V_1 \cup V_2, E)$ is a bipartite graph. Consider the bipartite graph $G' = (V_1' \cup V_2', E')$, where $V_1' = V_1 \cup \{a_1, b_1\}$, $V_2' = V_2 \cup \{a_2, b_2\}$, and $E' = E' \cup \{a_1 a_2, a_1 b_2, a_2 b_1\} \cup \{xa_2\}_{x \in V_1} \cup \{a_1 y\}_{y \in V_2}$ (see Fig. 7.1). Notice that $\operatorname{diam}(G') = 3$ and check that $g(G') = \gamma(G) + 2$. Finally, use the fact that the Domination Number Problem for bipartites graphs is NP-complete [145].

Assume now that $G = (V, E)$ is a chordal graph. Consider the chordal graph $G' = (V', E')$, where $V' = V \cup \{x_u, y_u\}_{u \in V} \cup \{z\}$ and $E' = E \cup \{ux_u, uy_u\}_{u \in V} \cup \{zu, zx_u\}_{u \in V}$. Notice that $\operatorname{diam}(G') = 4$ and check that $g(G') = \gamma(G) + |V|$. Finally, use the fact that the Domination Number Problem for chordal graphs is NP-complete [145]. □

Corollary 7.1 ([6,110,113]). *The problem of determining the geodetic number of a graph is an NP-hard problem with its recognition problem stated as an NP-complete problem.*

Given a graph G and a positive integer k, the problem of deciding whether $h(G) \leq k$ is called the *Hull Number Problem*.

Theorem 7.2 ([81]). *The Hull Number Problem is NP-complete.*

Given a graph G, if there exists a partition of $V(G)$ into p convex sets, we say that G is *p-convex*. The problem of deciding whether a graph G is p-convex for a fixed integer p is said to be the *Convex p-Partition Problem*. Similarly, if there

I.M. Pelayo, *Geodesic Convexity in Graphs*, SpringerBriefs in Mathematics,
DOI 10.1007/978-1-4614-8699-2_7, © Ignacio M. Pelayo 2013

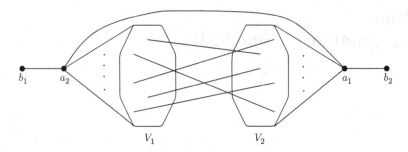

Fig. 7.1 Bipartite graph obtained from a given bipartite graph $G = (V_1 \cup V_2, E)$

exists a partition of $V(G)$ into p cliques we say that G is *p-clique*. The problem of deciding whether a graph G is p-clique for a fixed integer p is said to be the *Clique p-Partition Problem*.

Certainly, p-clique graph is p-convex graph, since every clique is convex. Notice also that unlike the clique case, the fact that G is p-convex does not implies that G is $p+1$-convex. For example, the graph G obtained by adding a new vertex and joining it to an arbitrary vertex of $K_{4,4}$ is both two-convex and four-convex, but it is not three-convex.

Theorem 7.3 ([4]). *The Convex p-Partition Problem is NP-complete.*

Sketch of proof. Let G be a nontrivial graph of diameter at least 2. Consider the graph G' obtained from G by adding two nonadjacent vertices u and v with $N_{G'}(u) = N_{G'}(v) = V(G)$. Check that any proper convex set of G' is a clique. Hence, if $V(G)$ admits a partition $\mathcal{V}$ into p cliques, then we can form a convex p-partition $\mathcal{V}'$ of $V(G')$ by adding u, v in different sets of $\mathcal{V}$. Conversely, a convex p-partition $\mathcal{V}'$ of $V(G')$ induces a partition of $V(G)$ into q cliques, where $p - 2 \leq l \leq q$. If $q \neq p$, divide a clique of $\mathcal{V}'$ into two cliques in order to obtain a partition into $q + 1$ cliques. If $q + 1 \neq p$, then repeat this argument once again.

Next, use the fact the Clique p-Partition Problem is NP-complete for $p \geq 3$ [131]. As for the case $p = 2$, the proof is based on a reduction from 1-in-3 SAT. $\square$

Given a graph G and a positive integer k, the problem of deciding whether $\text{con}(G) \geq k$ is called the *Convexity Number Problem*.

Theorem 7.4 ([104]). *The Convexity Number Problem is NP-complete.*

Proof. Given a graph G of order n, let H be the join of G with two isolated vertices, say u and v. Let K be a proper convex set of H. Suppose that there exists a pair x, y of distinct nonadjacent vertices in K. Clearly, $\{x, y\} \neq \{u, v\}$ as $I[u, v] = V(H)$. Observe also that both u and v belong to $x - y$ paths of length 2, which means that $u, v \in I[x, y]$, i.e., $V(H) = I[u, v] \subseteq I[x, y]$, a contradiction. Thus, the largest proper convex sets in H are complete graphs containing u or v. In other words, $\text{con}(H) = \omega(H) = \omega(G) + 1$. Finally, use the fact that the clique number problem is NP-complete [131]. $\square$

Theorem 7.5 ([3, 86]). *Both the Hull Number Problem and the Convexity Number Problem, restricted to bipartite graphs are NP-complete.*

Conjecture 7.1. The Convex p-Partition Problem restricted to bipartite graphs is NP-complete.

Theorem 7.6 ([4, 81, 84, 86]). *If G is a P_4-free graph, then the Geodetic Number, the Hull Number, the Convexity Number, and the Convex p-partition problems can be decided in linear time.*

Further results involving computational complexity problems related to geodesic convexity in graphs can be found in [3, 4, 6, 38, 80, 81, 83, 84, 86, 95, 104, 110, 113, 123, 140, 164].

Glossary

All-paths convexity A set $S \subseteq V(G)$ is convex if for any pair of vertices x, y of S, each path connecting x and y is completely contained in S.

Antihole Complement of a hole.

Antimatroid Convex geometry.

Bipartite graph Its vertex set can be partitioned into two independent sets.

Boundary Subgraph induced by the set of locally eccentric vertices.

Carathéodory number Maximum cardinality of an irredundant set.

Chord Edge joining two non-consecutive vertices of a cycle or a path.

Chordal graph It contains no induced cycles of length greater than three.

Chordless path convexity Monophonic convexity.

Clique An induced subgraph isomorphic to a complete graph.

Clique number Maximum order of a clique.

Closure operator A closure operator ϕ on a set V is a function $\phi : 2^V \mapsto 2^V$ which satisfies the following conditions for all sets $A, B \in 2^V$: (1) $A \subseteq \phi(A)$, (2) $A \subseteq B \Rightarrow \phi(A) \subseteq \phi(B)$ and (3) $\phi(\phi(A)) = A$.

Constructible graph There exists a well-ordering $\leq$ of its vertices such that, for every vertex x which is not the smallest element, there is a vertex $y < x$ which is adjacent to x and to every neighbor z of x with $z < x$.

Contour Subgraph induced by the set of locally peripheral vertices.

Convex geometry Convexity satisfying the Krein–Milman property: Every convex set is the convex hull of its extreme vertices.

Convex hull The convex hull $[S]$ of a set S is the smallest convex set containing S.

Convex set Given a graph convexity space (V,C) of a graph $G = (V,E)$, the elements of C are called the convex sets of G.

Convexity number Cardinality of a maximum proper convex set.

Convexity space A convexity space is a pair $(V,\mathscr{C})$ consisting of a finite set V and a set C of subsets of V such that $\emptyset \in \mathscr{C}$, $V \in \mathscr{C}$, $\mathscr{C}$ is closed under intersection and the union of every chain of elements of $\mathscr{C}$ is in $\mathscr{C}$.

Detour convexity Longest path convexity.

Distance-hereditary graph Every chordless path is a geodesic.

Extreme set Set of simplicial vertices.

Extreme subgraph Subgraph induced by the set of simplicial vertices.

Extreme vertex A vertex $x \in X$ is an extreme vertex of X if $X \setminus \{x\}$ is convex.

Eccentric subgraph Set of eccentric vertices.

Eccentric subgraph Subgraph induced by the set of eccentric vertices.

Eccentric vertex A vertex v is an eccentric vertex of another vertex u if no vertex is further away from u than v.

Eccentricity The eccentricity of a vertex v is the maximum distance between v and any other vertex.

g^3-convexity A set $S \subseteq V(G)$ is convex if for any pair of vertices x,y of S, each shortest path of length at least three connecting x and y is completely contained in S.

g_k-convexity A set $S \subseteq V(G)$ is convex if for any pair of vertices x,y of S, each shortest path of length at most k connecting x and y is completely contained in S.

Geodesic Shortest path.

Geodesic convexity A set $S \subseteq V(G)$ is g-convex if for any pair of vertices x,y of S, each shortest path connecting x and y is completely contained in S.

Geodesically convex set Any geodesic joining two vertices of this set lies entirely within it.

Geodetic closure The geodetic closure $I[S]$ of S is the union of all geodetic intervals $I[u,v]$ taken over all pairs $u,v \in S$.

Geodetic interval Set of all vertices belonging to a shortest $u - v$ path.

Geodetic number Minimum cardinality of a geodetic set.

Geodetic set A set S of vertices such that $I[S] = V(G)$.

Geodominating set Geodetic set.

HHD-free graph Graph in which every cycle of length at least 5 has at least 2 chords.

Helly independent set A set such that $\bigcap_{a \in A}[A - a] = \emptyset$.

Helly number Maximum cardinality of a Helly independent set.

Hole Cycle of length at leas five.

Hull number Minimum cardinality of a hull set.

Hull set A set S of vertices such that $[S] = V(G)$.

Hypercube Cartesian product of complete graphs of order 2.

Induced path convexity Monophonic convexity.

Interval graph Intersection graph of a set of intervals on the real line.

Irredundant set A set such that $\bigcup_{a \in A}[A - a] \subsetneq [A]$.

k-**set** Set of cardinality k.

k-**Steiner distance-hereditary graph** A graph G is k-SDH if for every connected induced subgraph H of G and every set S of k vertices of H, $d_H(S) = d_G(S)$.

Leaf End-vertex, i.e., vertex of degree 1.

Locally eccentric vertex A vertex v is a locally eccentric vertex of another vertex u if no neighbor of v is further away from u than v.

Locally peripheral vertex A vertex whose eccentricity is equal or greater than the maximum of the eccentricities of its neighbors.

Longest path convexity A set $S \subseteq V(G)$ is convex if for any pair of vertices x, y of S, each longest path connecting x and y is completely contained in S.

m-**convexity** Monophonic convexity.

m^3-**convexity** A set $S \subseteq V(G)$ is convex if for any pair of vertices x, y of S, each induced path of length at least three connecting x and y is completely contained in S.

Minimal path convexity Monophonic convexity.

Monophonic closure The monophonic closure $J[S]$ of S is the union of all monophonic intervals $J[u, v]$ taken over all pairs $u, v \in S$.

Monophonic convexity A set $S \subseteq V(G)$ is m-convex if for any pair of vertices x, y of S, each induced path connecting x and y is completely contained in S.

Monophonic interval Set of all vertices belonging to a chordless $u - v$ path.

P_3-**convexity** Two-path convexity.

Peripheral vertex A vertex whose eccentricity equals the diameter.

Periphery Subgraph induced by the set of peripheral vertices.

Ptolemaic graph Chordal graph which is distance-hereditary.

Radon number Maximum cardinality of a Radon independent set.

Radon independent set A set not admitting a Radon partition.

Radon partition Is a partition $\{A_1,A_2\}$ of a set $A \subseteq V$ of a convexity space $(V,\mathscr{C})$ such that $[A_1]_\mathscr{C} \cap [A - a]_\mathscr{C} \neq \emptyset$.

Shortest path An (x,y)-path of length $d(x,y)$.

Simplicial vertex The subgraph induced by its neighborhood is a clique.

Steiner interval The Steiner interval of a set of vertices W, is the set of all vertices that lie on some Steiner W-tree.

Steiner tree A Steiner tree of a set of vertices A is any connected subgraph that contains A and has the minimum number of edges.

Steiner convex set A set $W \subseteq V(G)$ is Steiner convex if for every $A \subseteq W, S[A] \subseteq W$.

Support vertex A vertex of a tree that is adjacent to a leaf of the tree.

Total convexity A set $S \subseteq V(G)$ is convex if for any pair of vertices x,y of S, each triangle free path connecting x and y is completely contained in S.

Triangle-free graph K_3-free graph, i.e., it contains no induced subgraph isomorphic to K_3.

Triangle-free path convexity Total convexity.

Triangle-path convexity A set $S \subseteq V(G)$ is convex if for any pair of vertices x,y of S, each triangle path connecting x and y is completely contained in S.

Two-path convexity A set $S \subseteq V(G)$ is convex if for any pair of vertices x,y of S, each path of length two connecting x and y is completely contained in S.

Weakly chordal graph A graph is weakly chordal if it does not contain neither holes nor antiholes as induced subgraphs.

Weakly modular graph A graph is weakly modular if for any triple $\{x,y,z\}$ of vertices such that $I[x,y] \cap I[x,z] = \{x\}, I[y,x] \cap I[y,z] = \{y\}$ and $I[z,x] \cap I[z,y] = \{z\}$, all vertices on every $y - z$ geodesic have the same distance to x.

References

1. Anand, B.S., Changat, M., Klavžar, S., Peterin, I.: Convex sets in lexicographic products of graphs. Graphs Combin. **28**, 77–84 (2012)
2. Aniversario, I.S., Jamil, F.P., Canoy, S.R.: The closed geodetic numbers of graphs. Util. Math. **74**, 3–18 (2007)
3. Araujo, J., Campos, V., Giroire, F., Nisse, N., Sampaio, L., Soares, R.: On the hull number of some graph classes. Theor. Comput. Sci. **475**, 1–12 (2013)
4. Artigas, D., Dantas, S., Dourado, M.C., Szwarcfiter, J.L.: Partitioning a graph into convex sets. Discrete Math. **311**(17), 1968–1977 (2011)
5. Artigas, D., Dantas, S., Dourado, M.C., Szwarcfiter, J.L., Yamaguchi, S.: On the contour of graphs. Discrete Appl. Math. **161**(10–11), 1356–1362 (2013)
6. Atici, M.: Computational complexity of geodetic set. Int. J. Comput. Math. **79**(5), 587–591 (2002)
7. Atici, M.: On the edge geodetic number of a graph. Int. J. Comput. Math. **80**(7), 853–861 (2003)
8. Atici, M., Vince, A.: Geodesics in graphs, an extremal set problem and perfect Hash families. Graphs Combin. **18**(3), 403–413 (2002)
9. Bandelt, H.-J., Chepoi, V.: A Helly theorem in weakly modular spaces. Discrete Math. **125**, 25–39 (1996)
10. Bandelt, H.-J., Chepoi, V.: Metric graph theory and geometry: a survey. In: Goodman, J.E., Pach, J., Pollack, R. (eds.) Surveys on Discrete and Computational Geometry: Twenty Years Later. Contemporary Mathematics, vol. 453, pp. 49–86. American Mathematical Society, Providence (2008)
11. Bandelt, H.-J., Mulder, H.M.: Distance-hereditary graphs. J. Comb. Theory Ser. B **41**(2), 182–208 (1986)
12. Barbosa, R.M., Coelho, E.M.M., Dourado, M.C., Rautenbach, D., Szwarcfiter, J.L.: On the Carathéodory number for the convexity of paths of order three. SIAM J. Discrete Math. **26**(3), 929–939 (2012)
13. Bermudo, S., Rodríguez-Velázquez, J.A., Sigarreta, J.M., Yero, I.G.: On geodetic and *k*-geodetic sets in graphs. Ars Combin. **96**, 469–478 (2010)
14. Brandstädt, A., V.B., Spinrad, J.: Graph classes: a survey. SIAM Monographs on Discrete Mathematics and Applications. SIAM, Philadelphia (1999)
15. Brešar, B., Gologranc, T.: On a local 3-Steiner convexity. European J. Combin. **32**(8), 1222–1235 (2011)
16. Brešar, B., Tepeh Horvat, A.: On the geodetic number of median graphs. Discrete Math. **308**(18), 4044–4051 (2008)

I.M. Pelayo, *Geodesic Convexity in Graphs*, SpringerBriefs in Mathematics,
DOI 10.1007/978-1-4614-8699-2, © Ignacio M. Pelayo 2013

17. Brešar, B., Klavžar, S., Tepeh Horvat, A.: On the geodetic number and related metric sets in Cartesian product graphs. Discrete Math. **308**(23), 5555–5561 (2008)

18. Brešar, B., Changat, M., Mathews, J., Peterin, I., Narasimha-Shenoi, P.G., Tepeh Horvat, A.: Steiner intervals, geodesic intervals, and betweenness. Discrete Math. **309**(20), 6114–6125 (2009)

19. Brešar, B., Kovše, M., Tepeh Horvat, A.: Geodetic sets in graphs. In: Structural Analysis of Complex Networks, pp. 197–218. Birkhäuser/Springer, New York (2011)

20. Brešar, B., Šumenjak, T.K., Tepeh Horvat, A.: The geodetic number of the lexicographic product of graphs. Discrete Math. **311**(16), 1693–1698 (2011)

21. Buckley, F., Harary, F.: Distance in Graphs. Addison-Wesley, Redwood City (1990)

22. Buckley, F., Harary, F., Quintas, L.V.: Extremal results on the geodetic number of a graph. Scientia **2A**, 17–26 (1988)

23. Cáceres, J., Márquez, A., Oellermann, O.R., Puertas, M.L.: Rebuilding convex sets in graphs. Discrete Math. **297**(1–3), 26–37 (2005)

24. Cáceres, J., Hernando, C., Mora, M., Pelayo, I.M., Puertas, M.L., Seara, C.: On geodetic sets formed by boundary vertices. Discrete Math. **306**(2), 188–198 (2006)

25. Cáceres, J., Hernando, C., Mora, M., Pelayo, I.M., Puertas, M.L., Seara, C.: Geodeticity of the contour of chordal graphs. Discrete Appl. Math. **156**(7), 1132–1142 (2008)

26. Cáceres, J., Márquez, A., Puertas, M.L.: Steiner distance and convexity in graphs. European J. Combin. **29**(3), 726–736 (2008)

27. Cáceres, J., Hernando, C., Mora, M., Pelayo, I.M., Puertas, M.L.: On the geodetic and the hull numbers in strong product graphs. Comput. Math. Appl. **60**(11), 3020–3031 (2010)

28. Cáceres, J., Oellermann, O.R., Puertas, M.L.: Minimal trees and monophonic convexity. Discuss. Math. Graph Theory **32**(4), 685–704 (2012)

29. Cagaanan, G.B., Canoy, S.R.: On the hull sets and hull number of the Cartesian product of graphs. Discrete Math. **287**, 141–144 (2004)

30. Cagaanan, G.B., Canoy, S.R.: On the geodetic covers and geodetic bases of the composition $G[K_m]$. Ars Combin. **79**, 33–45 (2006)

31. Cagaanan, G.B., Canoy, S.R.: Bounds for the geodetic number of the Cartesian product of graphs. Util. Math. **79**, 91–98 (2009)

32. Calder, J.R.: Some elementary properties of interval convexities. J. London Math. Soc. **3**(2), 422–428 (1971)

33. Canoy, S.R., Cagaanan, G.B.: On the geodesic and hull numbers of the sum of graphs. Congr. Numer. **161**, 97–104 (2003)

34. Canoy, S.R., Cagaanan, G.B.: On the hull number of the composition of graphs. Ars Combin. **75**, 113–119 (2005)

35. Canoy, S.R., Garces, I.J.L.: Convex sets under some graph operations. Graphs Combin. **18**(4), 787–793 (2002)

36. Canoy, S.R., Cagaanan, G.B., Gervacio, S.V.: Convexity, geodetic and hull numbers of the join of graphs. Util. Math. **71**, 143–159 (2006)

37. Centeno, C.C., Dantas, S., Dourado, M.C., Rautenbach, D., Szwarcfiter, J.L.: Convex partitions of graphs induced by paths of order three. Discrete Math. Theory Comput. Sci. **12**(5), 175–184 (2010)

38. Chae, G.-B., Palmer, E.M., Siu, W.-C.: Geodetic number of random graphs of diameter 2. Australas. J. Combin. **26**, 11–20 (2002)

39. Chang, G.J., Tong, L.-D., Wang, H.-T.: Geodetic spectra of graphs. European J. Combin. **25**(3), 383–391 (2004)

40. Changat, M., Mathew, J.: On triangle path convexity in graphs. Discrete Math. **206**, 91–95 (1999)

41. Changat, M., Mathew, J.: Induced path transit function, monotone and Peano axioms. Discrete Math. **286**(3), 185–194 (2004)

42. Changat, M., Klavžar, S., Mulder, H.M.: The all-paths transit function of a graph. Czechoslovak Math. J. **51**(2), 439–448 (2001)

43. Changat, M., Mulder, H.M., Sierksma, G.: Convexities related to path properties on graphs. Discrete Math. **290**(2–3), 117–131 (2005)
44. Changat, M., Mathew, J., Mulder, H.M.: The induced path function, monotonicity and betweenness. Discrete Appl. Math. **158**(5), 426–433 (2010)
45. Changat, M., Narasimha-Shenoi, P.G., Mathews, J.: Triangle path transit functions, betweenness and pseudo-modular graphs. Discrete Math. **309**(6), 1575–1583 (2009)
46. Changat, M., Narasimha-Shenoi, P.G., Pelayo, I.M.: The longest path transit function of a graph and betweenness. Util. Math. **82**, 111–127 (2010)
47. Changat, M., Lakshmikuttyamma, A.K., Mathews, J., Peterin, I., Prasanth, N.-S., Tepeh, A.: A note on 3-Steiner intervals and betweenness. Discrete Math. **311**(22), 2601–2609 (2011)
48. Chartrand, G., Zhang, P.: Convex sets in graphs. Congr. Numer. **136**, 19–32 (1999)
49. Chartrand, G., Zhang, P.: The forcing geodetic number of a graph. Discuss. Math. Graph Theory **19**, 45–58 (1999)
50. Chartrand, G., Zhang, P.: The geodetic number of an oriented graph. Europ. J. Combin. **21**, 181–189 (2000)
51. Chartrand, G., Zhang, P.: On graphs with a unique minimum hull set. Discuss. Math. Graph Theory **21**(1), 31–42 (2001)
52. Chartrand, G., Zhang, P.: The forcing convexity number of a graph. Czechoslovak Math. J. **51**(4), 847–858 (2001)
53. Chartrand, G., Zhang, P.: The forcing hull number of a graph. J. Combin. Math. Combin. Comput. **36**, 81–94 (2001)
54. Chartrand, G., Zhang, P.: Extreme Geodesic Graphs. Czechoslovak Math. J. **52**(4), 771–780 (2002)
55. Chartrand, G., Zhang, P.: The Steiner number of a graph. Discrete Math. **242**, 41–54 (2002)
56. Chartrand, G., Oellermann, O.R., Tian, S., Zou, H.: Steiner distance in graphs. Časopis Pěst. Mat. **114**(4), 399–410 (1989)
57. Chartrand, G., Zhang, P., Harary, F.: Extremal problems in geodetic graph theory. Cong. Numer. **131**, 55–66 (1998)
58. Chartrand, G., Harary, F., Zhang, P.: Geodetic sets in graphs. Discuss. Math. Graph Theory **20**, 129–138 (2000)
59. Chartrand, G., Harary, F., Zhang, P.: On the hull number of a graph. Ars Combin. **57**, 129–138 (2000)
60. Chartrand, G., Harary, F., Swart, H.C., Zhang, P.: Geodomination in graphs. Bull. Inst. Combin. Appl. **31**, 51–59 (2001)
61. Chartrand, G., Chichisan, A., Wall, C.E., Zhang, P.: On convexity in graphs. Congr. Numer. **148**, 33–41 (2001)
62. Chartrand, G., Fink, J.F., Zhang, P.: Convexity in oriented graphs. Discrete Math. **116**(1–2), 115–126 (2002)
63. Chartrand, G., Harary, F., Zhang, P.: On the geodetic number of a graph. Networks **39**, 1–6 (2002)
64. Chartrand, G., Wall, C.E., Zhang, P.: The convexity number of a graph. Graphs and Combin. **18**(2), 209–217 (2002)
65. Chartrand, G., Palmer, E.M., Zhang, P.: The geodetic number of a graph: a survey. Congr. Numer. **156**, 37–58 (2002)
66. Chartrand, G., Erwin, D., Johns, G.L., Zhang, P.: Boundary vertices in graphs. Discrete Math. **263**, 25–34 (2003)
67. Chartrand, G., Fink, J.F., Zhang, P.: The hull number of an oriented graph. Int. J. Math. Math. Sci. **36**, 2265–2275 (2003)
68. Chartrand, G., Garry, L., Zhang, P.: The detour number of a graph. Util. Math. **64**, 97–113 (2003)
69. Chartrand, G., Garry, L., Zhang, P.: On the detour number and geodetic number of a graph. Ars Combin. **72**, 3–15 (2004)
70. Chartrand, G., Escuadro, H., Zhang, P.: Detour distance in graphs. J. Combin. Math. Combin. Comput. **52**, 75–94 (2005)

71. Chartrand, G., Lesniak, L., Zhang, P.: Graphs and Digraphs, 5th edn. CRC Press, Boca Raton (2011)
72. Chastand, M., Polat, N.: On geodesic structures of weakly median graphs I. Decomposition and octahedral graphs. Discrete Math. **306**(13), 1272–1284 (2006)
73. Chastand, M., Polat, N.: On geodesic structures of weakly median graphs II: Compactness, the role of isometric rays. Discrete Math. **306**(16), 1846–1861 (2006)
74. Chepoi, V.: Isometric subgraphs of Hamming graphs and d-convexity. Cyber. Syst. Anal. **24**(1), 6–11 (1988)
75. Chepoi, V.: Separation of two convex sets in convexity structures. J. Geom. **50**(1–2), 30–51 (1994)
76. Chepoi, V.: Peakless functions on graphs. Discrete Appl. Math. **73**(2), 175–189 (1997)
77. Cyman, J., Lemanska, M., Raczek, J.: Graphs with convex domination number close to their order. Discuss. Math. Graph Theory **26**(2), 307–316 (2006)
78. Dirac, G.A.: On rigid circuit graphs. Abh. Math. Sem. Univ. Hamburg **25**, 71–76 (1961)
79. Dong, L., Lu, C., Wang, X.: The upper and lower geodetic numbers of graphs. Ars Combin. **91**, 401–409 (2009)
80. Dourado, M.C., Protti, F., Szwarcfiter, J.L.: On the complexity of the geodetic and convexity numbers of a graph. RMS Lect. Notes Ser. Math. **7**, 101–108 (2008)
81. Dourado, M.C., Gimbel, J.G., Kratochvíl, J., Protti, F., Szwarcfiter, J.L.: On the computation of the hull number of a graph. Discrete Math. **309**(18), 5668–5674 (2009)
82. Dourado, M.C., Protti, F., Rautenbach, D., Szwarcfiter, J.L.: On the hull number of triangle-free graphs. SIAM J. Discrete Math. **23**(4), 2163–2172 (2009/10)
83. Dourado, M.C., Protti, F., Szwarcfiter, J.L.: Complexity results related to monophonic convexity. Discrete Appl. Math. **158**(12), 1268–1274 (2010)
84. Dourado, M.C., Protti, F., Rautenbach, D., Szwarcfiter, J.L.: Some remarks on the geodetic number of a graph. Discrete Math. **310**, 832–837 (2010)
85. Dourado, M.C., Rautenbach, D., dos Santos, V.F., Schäfer, P.M., Szwarcfiter, J.L., Toman, A.: An upper bound on the P_3-Radon number. Discrete Math. **312**(16), 2433–2437 (2012)
86. Dourado, M.C., Protti, F., Rautenbach, D., Szwarcfiter, J.L.: On the convexity number of graphs. Graphs Combin. **28**(3), 333–345 (2012)
87. Dourado, M.C., Rautenbach, D., de Sá, V.G.P., Szwarcfiter, J.L.: On the geodetic radon number of grids. Discrete Math. **313**(1), 111–121 (2013)
88. Dragan, F.F., Nicolai, F., Brandstädt, A.: Convexity and HHD-free graphs. SIAM J. Discrete Math. **12**, 119–135 (1999)
89. Duchet, P.: Convex sets in graphs II. Minimal path convexity. J. Comb. Theory Ser. B **44**(3), 307–316 (1988)
90. Duchet, P.: Convexity in combinatorial structures. Rend. Circ. Mat. Palermo **14**(2), 261–293 (1987)
91. Duchet, P.: Discrete convexity: retractions, morphisms and the partition problem. Proceedings of the Conference on Graph Connections, India, 1998, pp. 10–18. Allied Publishers, New Delhi (1999)
92. Duchet, P.: Radon and Helly numbers of segment spaces. Ramanujan Math. Soc. Lect. Notes Ser. **5**, 57–71 (2008)
93. Duchet, P., Meyniel, H.: Convex sets in graphs I. Helly and Radon theorems for graphs and surfaces. European J. Combin. **4**(2), 127–132 (1983)
94. Edelman, P.H., Jamison, R.E.: The theory of convex geometries. Geometriae Dedicata **19**, 247–270 (1985)
95. Ekim, T., Erey, A., Heggernes, P., van't Hof, P., Mesiter , D.: Computing minimum geodetic sets in proper interval graphs. Lect. Notes Comput. Sci. **7256**, 279–290 (2012)
96. Eroh, L., Oellermann, O.R.: Geodetic and Steiner sets in 3-Steiner distance hereditary graphs. Discrete Math. **308**(18), 4212–4220 (2008)
97. Escuadro, H., Gera, R., Hansberg, A., Rad, N.J., Volkmann, L.: Geodetic domination in graphs. J. Combin. Math. Combin. Comput. **77**, 89–101 (2011)
98. Everett, M.G., Seidman, S.B.: The hull number of a graph. Discrete Math. **57**, 217–223 (1985)

99. Farber, M.: Bridged graphs and geodesic convexity. Discrete Math. **66**, 249–257 (1987)

100. Farber, M., Jamison, R.E.: Convexity in graphs and hypergraphs. SIAM J. Alg. Disc. Math. **7**(3), 433–444 (1986)

101. Farber, M., Jamison, R.E.: On local convexity in graphs. Discrete Math. **66**, 231–247 (1987)

102. Farrugia, A.: Orientable convexity, geodetic and hull numbers in graphs. Discrete Appl. Math. **148**(3), 256–262 (2005)

103. Frucht, R., Harary, F.: On the corona of two graphs. Aequationes Math. **4**, 322–325 (1970)

104. Gimbel, J.G.: Some remarks on the convexity number of a graph. Graphs Combin. **19**, 357–361 (2003)

105. Goddard, W.: A note on Steiner-distance-hereditary graphs. J. Combin. Math. Combin. Comput. **40**, 167–170 (2002)

106. Gruber, P.M., Wills, J.M.: Handbook of Convex Geometry, (v. A-B). North-Holland, Amsterdam (1993)

107. Gutin, G., Yeo, A.: On the number of connected convex subgraphs of a connected acyclic digraph. Discrete Appl. Math. **157**(7), 1660–1662 (2009)

108. Hammark, R., Imrich, R., Klavžar, S.: Handbook of Product Graphs. CRC Press, Boca Raton (2011)

109. Hansberg, A., Volkmann, L.: On the geodetic and geodetic domination numbers of a graph. Discrete Math. **310**(15–16), 2140–2146 (2010)

110. Hansen, P., van Omme, N.: On pitfalls in computing the geodetic number of a graph. Optim. Lett. **1**(3), 299–307 (2007)

111. Harary, F.: Achievement and avoidance games for graphs. Ann. Discrete Math. **13**(11), 111–119 (1982)

112. Harary, F., Nieminen, J.: Convexity in graphs. J. Differential Geom. **16**(2), 185–190 (1981)

113. Harary, F., Loukakis, E., Tsouros, C.: The geodetic number of a graph. Math. Comput. Modelling **17**(11), 89–95 (1993)

114. Hasegawa, Y., Saito, A.: Graphs with small boundary. Discrete Math. **307**(14), 1801–1807 (2007)

115. Henning, M.A., Nielsen, M.H., Oellermann, O.R.: Local Steiner convexity. European J. Combin. **30**, 1186–1193 (2009)

116. Hernando, C., Jiang, T., Mora, M., Pelayo, I.M., Seara, C.: On the Steiner, geodetic and hull numbers of graphs. Discrete Math. **293**(1–3), 139–154 (2005)

117. Hernando, C., Mora, M., Pelayo, I.M., Seara, C.: Some structural, metric and convex properties on the boundary of a graph. Electron. Notes Discrete Math. **24**, 203–209 (2006)

118. Hernando, C., Mora, M., Pelayo, I.M., Seara, C.: Some structural, metric and convex properties of the boundary of a graph. Ars Combin. **109**, 267–283 (2013)

119. Hernando, C., Mora, M., Pelayo, I.M., Seara, C.: On monophonic sets in graphs. Submitted.

120. Howorka, E.: A characterization of distance-hereditary graphs. Quart. J. Math. Oxford Ser. 2 **28**(112), 417–420 (1977)

121. Howorka, E.: A characterization of Ptolemaic graphs. J. Graph Theory **5**(3), 323–331 (1981)

122. Hung, J.-T., Tong, L.-D., Wang, H.-T.: The hull and geodetic numbers of orientations of graphs. Discrete Math. **309**(8), 2134–2139 (2009)

123. Imrich, W., Klavžar, S.: A convexity lemma and expansion procedures for bipartite graphs. European J. Combin. **19**(6), 677–685 (1998)

124. Isaksen, D.C., Robinson, B.: Triangle-free polyconvex graphs. Ars Combin. **64**, 259–263 (2002)

125. Jamil, F.P.; Aniversario, I.S., Canoy, S.R.: On closed and upper closed geodetic numbers of graphs. Ars Combin. **84**, 191–203 (2007)

126. Jamil, F.P.; Aniversario, I.S., Canoy, S.R.: The closed geodetic numbers of the corona and composition of graphs. Util. Math. **82**, 135–153 (2010)

127. Jamison, R.E.: Copoints in antimatroids. Congr. Numer. **29**, 535–544 (1980)

128. Jamison, R.E.: Partition numbers for trees and ordered sets. Pacific J. Math. **96**, 115–140 (1981)

129. Jamison, R.E., Nowakowsky, R.: A Helly theorem for convexity in graphs. Disc. Math. **51**, 35–39 (1984)
130. Jiang, T., Pelayo, I.M., Pritikin, D.: Geodesic Convexity and Cartesian Products in Graphs. Manuscript (2004)
131. Karp, R.M.: Reducibility among combinatorial problems. In Miller, R.E., Thatcher, J.W. (eds.) Complexity of Computer Computations, pp. 85–103. Plenum, New York (1972)
132. Kay, D.C., Womble, E.W.: Axiomatic convexity theory and relationships between the Carathéodory, Helly and Radon numbers. Pacific J. Math. **38**, 471–485 (1971)
133. Klee, V.: What is a convex set. Am. Math. Mon. **78**, 616–631 (1971)
134. Korte, B., Lovász, L.: Homomorphisms and Ramsey properties of antimatroids. Discrete Appl. Math. **15** (2–3), 283–290 (1986)
135. Korte, B., Lovász, L., Schrader, R.: Gredoids. Springer, Berlin (1991)
136. Kubicka, E., Kubicki, G., Oellermann, O.R.: Steiner intervals in graphs. Discrete Appl. Math. **81**(1–3), 181–190 (1998)
137. Lemanska, M.: Weakly convex and convex domination numbers. Opuscula Math. **24**(2), 181–188 (2004)
138. Levi, F.W.: On Helly's theorem and the axioms of convexity. J. Indian Math. Soc. **15**, 65–76 (1951)
139. Lu, C.: The geodetic numbers of graphs and digraphs. Sci. China Ser. A: Math. **50**(8), 1163–1172 (2007)
140. Malvestuto, F.M., Mezzini, M., Moscarini, M.: Characteristic properties and recognition of graphs in which geodesic and monophonic convexities are equivalent. Discrete Math. Algorithms Appl. **4**(4), 1250063, pp. 14 (2012)
141. McKee, T.A., McMorris, F.R.: Topics in intersection graph theory. SIAM Monographs in Discrete Mathematics and Applications. SIAM, Philadelphia (1999)
142. Morgana, M.A., Mulder, H.M.: The induced path convexity, betweenness, and svelte graphs. Discrete Math. **254**(1–3), 349–370 (2002)
143. Mulder, H.M.: The interval function of a graph. Mathematical Centre Tracts. Mathematisch Centrum, Amsterdam (1980)
144. Mulder, H.M., Nebeský, L.: Axiomatic characterization of the interval function of a graph. European J. Combin. **30**(5), 1172–1185 (2009)
145. Müller, H., Brandstädt, A.: The NP-completeness of Steiner tree and dominating set for chordal bipartite graphs. Theoret. Comput. Sci. **53**(2–3), 257–265 (1987)
146. Muntean, R., Zhang, P.: On geodomination in graphs. Congr. Numer. **143**, 161–174 (2000)
147. Muntean, R., Zhang, P.: k-geodomination in graphs. Ars Combin. **63**, 33–47 (2002)
148. Nebeský, L.: A characterization of the interval function of a connected graph. Czechoslovak Math. J. **44**(1), 173–178 (1994)
149. Nielsen, M.H., Oellermann, O.R.: Helly theorems for 3-Steiner and 3-monophonic convexity in graphs. Discrete Math. **311**(10–11), 872–880 (2011)
150. Nielsen, M.H., Oellermann, O.R.: Steiner trees and convex geometries. SIAM J. Discrete Math. **23**(2), 680–693 (2011)
151. Oellermann, O.R., Peters-Fransen, J.: The strong metric dimension of graphs and digraphs. Discrete Appl. Math. **155**, 356–364 (2007)
152. Oellermann, O.R., Puertas, M.L.: Steiner intervals and Steiner geodetic numbers in distance-hereditary graphs. Discrete Math. **307**(1), 88–96 (2007)
153. Parker, D.B., Westhoff, R.F., Wolf, M.J.: Two-path convexity in clone-free regular multipartite tournaments. Australas. J. Combin. **36**, 177–196 (2006)
154. Parvathy, K.S., Vijayakumar, A.: Geodesic iteration number. Proceedings of the Conference on Graph Connections, India, 1998, pp. 91–94. Allied Publishers, New Delhi (1999)
155. Pelayo, I.M.: Comment on "The Steiner number of a graph" by G. Chartrand and P. Zhang [Discrete Math. **242**, 41–54 (2002)]. Discrete Math. **280**, 259–263 (2004)
156. Pelayo, I.M.: Generalizing the Krein-Milman property in graph convexity spaces: a short survey. Ramanujan Math. Soc. Lect. Notes Ser. **5**, 131–142 (2008)

157. Peterin, I.: The pre-hull number and lexicographic product. Discrete Math. **312**(14), 2153–2157 (2012)
158. Peterin, I.: Intervals and convex sets in strong product of graphs. Graphs Combin. **29**(3), 705–714 (2013)
159. Polat, N.: A Helly theorem for geodesic convexity in strongly dismantlable graphs. Discrete Math. **140**, 119–127 (1995)
160. Polat, N.: Graphs without isometric rays and invariant subgraph properties. I. J. Graph Theory **27**(2), 99–109 (1998)
161. Polat, N.: On isometric subgraphs of infinite bridged graphs and geodesic convexity. Discrete Math. **244**(1–3), 399–416 (2002)
162. Polat, N.: On constructible graphs, locally Helly graphs, and convexity. J. Graph Theory **43**(4), 280–298 (2003)
163. Polat, N., Sabidussi, G.: On the geodesic pre-hull number of a graph. European J. Combin. **30**(5), 1205–1220 (2009)
164. Raczek, J.: NP-completeness of weakly convex and convex dominating set decision problems. Opuscula Math. **24**(2), 189–196 (2004)
165. Sampathkumar, E.: Convex sets in a graph. Indian J. Pure Appl. Math. **15**(10), 1065–1071 (1984)
166. Santhakumaran, A.P., John, J.: Edge geodetic number of a graph. J. Discrete Math. Sci. Cryptogr. **10**(3), 415–432 (2007)
167. Sierksma, G.: Carathéodory and Helly-numbers of convex product structures. Pacific J. Math. **61**, 275–282 (1975)
168. Sierksma, G.: Relationships between Carathéodory, Helly, Radon and exchange numbers of convexity spaces. Nieuw Archief voor Wisk. **25**(2), 115–132 (1977)
169. Soltan, V.P.: Metric convexity in graphs. Studia Univ. Babeş-Bolyai Math. **36**(4), 3–43 (1991)
170. Soltan, V., Chepoi, V.: Conditions for invariance of set diameter under d-convexification in a graph. Cyber. Syst. Anal. **19**(6), 750–756 (1983)
171. Tong, L.-D.: The forcing hull and forcing geodetic numbers of graphs. Discrete Appl. Math. **157**(5), 1159–1163 (2009)
172. Tong, L.-D.: Geodetic sets and Steiner sets in graphs. Discrete Math. **309**(12), 4205–4207 (2009)
173. Tong, L.-D.: The (a, b)-forcing geodetic graphs. Discrete Math. **309**(6), 1623–1628 (2009)
174. Tong, L.-D., Yen, P.-L., Farrugia, A.: The convexity spectra of graphs. Discrete Appl. Math. **156**(10), 1838–1845 (2008)
175. Van de Vel, M.: Theory of Convex Structures. North-Holland, Amsterdam (1993)
176. Wang, F.-H.: The lower and upper forcing geodetic numbers of complete n-partite graphs, n-dimensional meshes and tori. Int. J. Comput. Math. **87**(12), 2677–2687 (2010)
177. Wang, F.-H., Wang, Y.-L., Chang, J.-M.: The lower and upper forcing geodetic numbers of block: cactus graphs. Eur. J. Oper. Res. **175**(1), 238–245 (2006)
178. Ye, Y., Lu, C., Liu, Q.: The geodetic numbers of cartesian products of graphs. Math. Appl. (Wuhan) **20**(1), 158–163 (2007)
179. Yen, P.-L.: A study of convexity spectra in directed graphs. Doctorate Dissertation, National Sun Yat-sen University, Taywan (2011)
180. Yero, I.G., Rodríguez-Velázquez, J.A.: Analogies between the geodetic number and the Steiner number of some classes of graphs. submitted
181. Zhang, P.: The upper forcing geodetic number of a graph. Ars Combin. **62**, 3–15 (2002)

Index

I.M. Pelayo, *Geodesic Convexity in Graphs*, SpringerBriefs in Mathematics,
DOI 10.1007/978-1-4614-8699-2, © Ignacio M. Pelayo 2013

Symbol Index

I.M. Pelayo, *Geodesic Convexity in Graphs*, SpringerBriefs in Mathematics, DOI 10.1007/978-1-4614-8699-2, © Ignacio M. Pelayo 2013